AF878681

Geoheritage, Geoparks and Geotourism

Conservation and Management Series

Series editors

Wolfgang Eder, Munich, Germany
Peter T. Bobrowsky, Burnaby, BC, Canada
Jesús Martínez-Frías, Madrid, Spain

Spectacular geo-morphological landscapes and regions with special geological features or mining sites, are becoming increasingly recognized as critical areas to protect and conserve for the unique geoscientific aspects they represent and as places to enjoy and learn about the science and history of our planet. More and more national and international stakeholders are engaged in projects related to "Geoheritage", "Geo-conservation", "Geoparks" and "Geo-tourism" and are positively influencing the general perception of modern Earth sciences. Most notably, "Geoparks", have proven to be excellent tools to educate the public about "Earth Sciences". And shown to be areas for recreation and significant sustainable economic development through geotourism. In order to develop further the understanding of earth sciences in general and to elucidate the importance of earth sciences for Society the Geoheritage, Geoparks and Geotourism Conservation and Management Series has been launched together with its sister GeoGuides series. 'Projects' developed in partnership with UNESCO, World Heritage and Global Geoparks Networks, IUGS and IGU, as well as with the 'Earth Science Matters' Foundation, are welcome. The series aims to provide a place for in-depth presentations of developmental and management issues related to Geoheritage and Geotourism as well existing and potential Geoparks. Individually authored monographs as well as edited volumes and conference proceedings are welcome in this series. This book series is considered to be complementary to the Springer-Journal "Geoheritage".

More information about this series at http://www.springer.com/series/11639

Mohammed Rashad Moufti
Károly Németh

Geoheritage of Volcanic Harrats in Saudi Arabia

Springer

Mohammed Rashad Moufti
Jeddah
Saudi Arabia

Károly Németh
Palmerston North
New Zealand

ISSN 2363-765X ISSN 2363-7668 (electronic)
Geoheritage, Geoparks and Geotourism
ISBN 978-3-319-33013-6 ISBN 978-3-319-33015-0 (eBook)
DOI 10.1007/978-3-319-33015-0

Library of Congress Control Number: 2016938386

© Springer International Publishing Switzerland 2016
This work is subject to copyright. All rights are reserved by the Publisher, whether the whole or part of the material is concerned, specifically the rights of translation, reprinting, reuse of illustrations, recitation, broadcasting, reproduction on microfilms or in any other physical way, and transmission or information storage and retrieval, electronic adaptation, computer software, or by similar or dissimilar methodology now known or hereafter developed.
The use of general descriptive names, registered names, trademarks, service marks, etc. in this publication does not imply, even in the absence of a specific statement, that such names are exempt from the relevant protective laws and regulations and therefore free for general use.
The publisher, the authors and the editors are safe to assume that the advice and information in this book are believed to be true and accurate at the date of publication. Neither the publisher nor the authors or the editors give a warranty, express or implied, with respect to the material contained herein or for any errors or omissions that may have been made.

Printed on acid-free paper

This Springer imprint is published by Springer Nature
The registered company is Springer International Publishing AG Switzerland

This book is dedicated to

His Royal Highness Prince Sultan Bin Salman Bin Abdulaziz, President of Saudi Commission for Tourism and National Heritage (SCTH)

and

His Royal Highness Prince Faisal Bin Salman Bin Abdulaziz, The Governor of Madinah Province, Kingdom of Saudi Arabia

Preface

This book is the first systematic summary of the volcanic geoheritage of the western margin of the Arabian Peninsula. In general volcanism is not commonly associated with the Arabian Peninsula, and rarely linked to the Kingdom of Saudi Arabia as being a country that is rich in volcanic sites worth to visit. This book hopefully will change this preconception as it will take the reader to a volcanic wonderland that can be traced over thousands of kilometres length in a several hundreds of kilometres wide belt that is aligned parallel with the present day coastline of the Red Sea. While this book provides a summary of the Cenozoic to Recent volcanic features located in the territory of the Kingdom of Saudi Arabia, these volcanoes are part of a much larger volcanic system that is probably the largest intracontinental volcanic province on Earth stretching from the southern regions of Eastern Turkey through Syria, Jordan from the north across the Kingdom of Saudi Arabia and ending in Yemen in the south. In this aspect this intracontinental volcanic province is comparable in size to the region commonly referred in the mainstream scientific literature as being the largest Cenozoic intracontinental volcanic region on Earth in eastern Australia. Especially the western regions of the Kingdom of Saudi Arabia host numerous volcanic fields that are called traditionally as harrats or volcanic lava fields—a synonym for volcanic fields—largely unknown to many of the global geological community. As these volcanic fields are located in arid climate and the majority of them were formed in the last 10 millions of years with several Quaternary eruption sites, the volcanic landforms are preserved exceptionally well. This book intend to provide the first systematic inventory of the volcanic geoheritage values of these unknown sites with an aim to provide scientific basis for future research and utilization of the geoheritage values of these volcanic regions.

The core of the book is a detailed description of the Harrat Rahat that is one of the largest volcanic fields in Saudi Arabia that hosts two well-documented historic eruption sites with well-preserved volcanic landforms with high geodiversity. The fresh volcanic appearance of Harrat Rahat is evident for any visitor. The region is also located in a culturally and religiously important triangle between Al Madinah, Makkah and Jeddah. As this region hosts one of the youngest volcanic eruption sites in the western Arabian Peninsula, this area got strong scientific attention from the early 1960s till today, hence a wealth of scientific knowledge were available to evaluate and catalogue the volcanic geoheritage values of the field. In addition since 2011, an international collaboration project between the King Abdulaziz University, Auckland University and Massey University focused on the Volcanic Risks in Saudi Arabia (VORiSA) and conducted several new researches to understand the volcanism in this region. This period also provided numerous opportunities to access volcanic sites that have not been studied before making a strong scientific basis to evaluate geoheritage values of this region realistically. This project leads for the first time to consider the northern part of Harrat Rahat, formally called as Harrat Al Madinah to be considered as the first volcanic geopark in the Kingdom of Saudi Arabia. In this respect the volcanic geoheritage values of Harrat Rahat became the backbone of this book.

In addition in this book we tried to sum up the current knowledge of the volcanic geoheritage of other regions in the Kingdom of Saudi Arabia. As the size of the region and their individual harrats are huge, and many of them have rarely been visited in the past, it was not an aim during the preparation of this book to be able to provide a well-balanced and in-depth inventory for every volcanic regions of the Kingdom of Saudi Arabia. We were focusing only on those areas where at least a single field campaign was arranged since 2011 and there were enough external data to be able to evaluate preliminarily the volcanic geoheritage values of those regions. As building a volcanic geoheritage inventory is an ongoing process, we hope that this book will provide a strong head start to demonstrate the high volcanic geodiversity the Kingdom of Saudi Arabia.

This book starts with an introduction where basic concepts of geoheritage studies are outlined with various definitions especially the way how these concepts and definitions are used throughout the book. A geological setting provides a simple but clear summary of the geological context of the Cenozoic volcanism of the Kingdom of Saudi Arabia. The main part of the book provides a step-by-step summary of the main geoheritage concepts of the main volcanic harrats of the region. Each chapter provides a summary of the basic concepts the volcanic geoheritage values of the specific harrats define prior a detailed inventory style summary of individual geosites and geotopes are provided. The book consists of two main parts; one of them is dedicated to Harrat Rahat, while the other chapter summarizes the volcanic geoheritage values of the Harrat Khaybar, Harrat Kishb, Harrat Hutaymah and Harrat al Birk. These are the harrats that have been studied from volcanic geoheritage perspective so far. Other harrats such as those located close to the northern regions close to the Syrian and Iraqi border to the Kingdom has not been studied in detail and currently accessing them is not easy. Other important harrats such as the Harrat Lunayyir which hosted a volcano-seismic unrest in 2009 has not been studied so far, as prior the event in 2009, the region was considered a fairly remote and unknown volcanic region that has not been included in volcanic geoheritage studies recently.

Overall we offer this book to anyone who would like to expand their horizons on volcanic regions so far been hidden from many people's eyes. This book can be used as a guide to locate the sites and plan geotourism or provide it for tour operators who can use it to design their own geotouristic and geoeducational programmes. This book also can be very useful for other geoheritage researchers either in the Kingdom of Saudi Arabia or elsewhere to follow techniques on how to create an inventory for volcanic geoheritage values of a region. The book can be a great help to develop future geoparks in the region as well as use the identified geotopes and geosites to scale and evaluate geopark projects elsewhere. We also hope that this book will be a main push to initiate a concentrated geoheritage work on the region to be able to propose the western Saudi Arabian intracontinental volcanic province as a unique and the largest volcanic province in intracontinental settings on Earth for the UNESCO World Heritage status.

Finally we would like to thank many people who helped to have this book published through excellent discussions over volcanic geoheritage values of volcanic fields, direct field missions and excellent review comments on various versions of the manuscript. Many thanks for Mohamed Abdelwahed, Essam Aboud, Faisal Alqahtani, Abdulrahman Alsahafi, Kate Arentsen, Peter Bitschene, Jose Brihla, Jess Cherrington, Shane Cronin, Nabil El-Masry, Ingomar Fritz, Kurt Goth, Catherine Kenedi, Gábor Kereszturi, Jan Lindsay, Volker Lorenz, Herbert Lutz, Hugo Murcia, Cecile Olive, Atef Qaddah, Ian Smith, Peter Suhr and Benjamin van Wyk de Vries. The field work of geoheritage researches was supported by the King Abdulaziz University (KAU) from Jeddah. This book is also the direct output of Károly Németh's sabbatical research stay in Jeddah, which was supported by the Geohazards Research Center at KAU and the Massey University.

Jeddah
Palmerston North

Mohammed Rashad Moufti
Károly Németh

Contents

1 Introduction

Geoconservation and geoeducation projects are increasingly popular in many volcanic fields on Earth (Doniz-Paez et al. 2011; El Hadi et al. 2011; Erfurt-Cooper 2011; Ghazi et al. 2013; Guijon et al. 2011; Henriques and Neto 2015; Kwon 2013; Moufti and Nemeth 2013; Moufti et al. 2013a, b; Sang 2014). Such projects not only ensure that future generations can visit protected and preserved volcanic geoheritage sites but also ensure that scientific focus will concentrate on understanding their geological value, especially from a volcanic hazard education point of view (Erfurt-Cooper 2011, 2014). One of the first steps in developing a suitable and sustainable volcanic geoheritage site is to establish a comprehensive database of potential geosites that can later be grouped and promoted in various scientific and educational programs under the umbrella of regional and global geoparks (Ahluwalia 2006; Armiero et al. 2011; Bruno et al. 2014; Hassan et al. 2012; Hwan 2011; Joyce 2010; Sheth et al. 2010; Strba et al. 2015; Tefogoum et al. 2014; Wang et al. 2014; Yildirim and Kocan 2008; 박종관 2013; 배수경 et al. 2014; 조규성 et al. 2014; 진광민 and Kim 2010). It is clear, as stated in various reports, that research on geoheritage sites, highlighting their regional and global value, is an emerging geoscience, which is gaining greater respect from the general end-users, educators and researchers (Erfurt-Cooper 2014; Errami et al. 2015; Woo et al. 2013). Identifying and recording the geological heritage of geosites and/or geomorphosites is based not only the pure scientific value of the site, but also its regional importance in terms of understanding the region's geological and geomorphological evolution (Guijón et al. 2011; Hassan et al. 2012; Corvea et al. 2014; Panizza 2001; Petrovic et al. 2013; Pica et al. 2014; Rocha et al. 2014; Ruban 2010; Strba et al. 2015; Tomic and Bozic 2014; Vujičić et al. 2011). This process is fundamentally driven by (1) the experts' categorisation of the diversity of a geosite and geomorphosite and (2) the preconception and prevailing attitude of the local population to the geodiversity and geoconservation value of the specific site (Brocx and Semeniuk 2007). This two-way approach to identify geodiversity and therefore establish a sustainable program to promote specific geosites, which can form the basis of regional and global geoeducation and geoconservation projects including the establishment of geoparks (both regional and global), is naturally a complex process (Ahluwalia 2006; Avram and Zarrilli 2012; Cardoso and Batista 2013; Godoy et al. 2013; Semeniuk et al. 2000). Geodiversity, therefore, can be a relative definition and could have very different meanings for the experts and the public (Comanescu and Nedelea 2012; Deraman et al. 2010; Gordon 2012; Gray 2008; Henriques et al. 2012; Lim 2014). To define appropriately a geosite's or geomorphosite's geological value certainly requires the geological and geomorphological expertise of scientists, while to make such geosites truly embedded in the life of a local community requires an understanding of the importance of the specific sites in the eyes of the local communities. This means that to successfully initiate a geoconservation, geoeducation and geotouristic project involves not only the promotion of the scientific value of the respective sites, but also investigation of the local community's view on the selected sites. However, it is clear that there are geological and geomorphological features that are so strikingly different from their surrounding areas that the gap between the scientific appraisal of their value and the prevailing view of the local community is likely to be little (Erfurt-Cooper 2011; Gordon 2011; Henriques et al. 2012; Moreira 2012; Reynard et al. 2011; Sanchez Cortez 2013). Undoubtedly such locations are the best places to initiate a geoconservation, geoeducation and geotouristic project.

Saudi Arabia is among those regions on Earth where the geological and geomorphological features are so diverse and so dramatic that to rank them in a hierarchical system is a difficult exercise. The region's arid climate and limited, exotic vegetation can make all the geological and geomorphological features very visible, due to their exposed nature, in a way that is rare in other regions under more humid climatic conditions. The dramatic weathering and erosional processes that can produce mass material over large areas that are seemingly stable over a long period of time can

© Springer International Publishing Switzerland 2016

M.R. Moufti and K. Németh, *Geoheritage of Volcanic Harrats in Saudi Arabia*,
Geoheritage, Geoparks and Geotourism, DOI 10.1007/978-3-319-33015-0_1

captivate the mind of the visitor with views and scales rare elsewhere.

In the western region of Saudi Arabia, a chain of volcanic ranges, locally referred to as harrats, form a very characteristic geomorphological feature that can be traced over 2000 km (Fig. 1.1). The majority of these lava-dominated landforms formed in the past 10 million years, and known historic volcanic eruptions made the landscape alive and influenced human society. The volcanic regions of western Saudi Arabia are unique due to their scientific value for understanding dispersed volcanism forming volcanic fields; their aesthetics and the overwhelming number of volcanoes that exist in the area; and their strong influence on the living nature that developed over them, including the human society that has flourished in this region for thousands of years. Overall, the harrats of Saudi Arabia can be viewed as forming a special region, with a high geoheritage value (Fig. 1.2). In this book we present an overview of the reason why intracontinental volcanism occurred in this part of Saudi Arabia, evaluate the region's geoheritage value and provide a strategy for how this region could be utilized to provide vital information to the scientific community and general

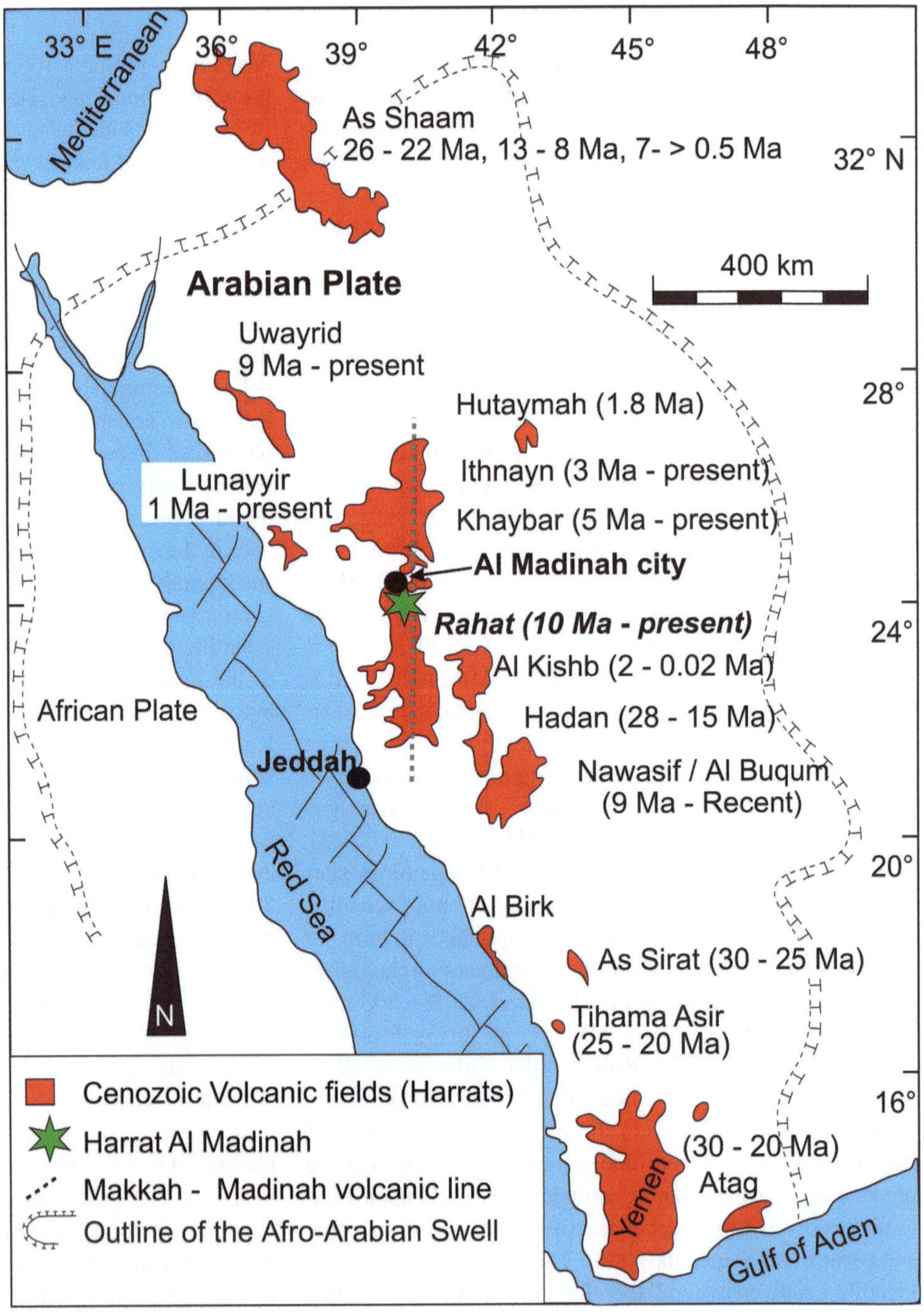

Fig. 1.1 Saudi Arabia's volcanic regions

Fig. 1.2 General overview of a harrat as a fantastic volcanic landscape of Western Saudi Arabia exhibiting scoria cones and extensive dark basaltic lava fields [1256 AD historic eruption site near Al Madinah city—photo taken from 24° 21′ 21.05″N; 39° 46′ 55.11″E toward west]

public to further understanding of eruption mechanisms and the history of dispersed volcanism.

The book also reflects the current state of geoheritage research in Saudi Arabia. The majority of geoheritage studies have focussed on the Al Madinah region in the northern Harrat Rahat, commonly referred to as Harrat Al Madinah (Fig. 1.2). A detailed description of the Harrat Al Madinah will form the backbone of this book, providing detailed information at the geosite level. The descriptions of the geosites will be accompanied by photo-documentation to justify the global significance of Harrat Al Madinah in understanding dispersed intracontinental volcanism. In the following chapters, conceptual models of the geoheritage values of other volcanic regions of Saudi Arabia will be given (Fig. 1.3). Those areas are equally important as the Harrat Al Madinah; however, the geoheritage research in those regions is not at a level yet where geosite-level inventories are able to be given. Moreover those other volcanic regions are so large that such research would provide work for many in the coming decades. Nevertheless, at this stage we can provide a framework for how our current understanding of their geoheritage values could be ordered into a logical and geoeducationally correct system. This could later serve as the scientific basis for the development of an Arabia-wide geopark network that could be a strong candidate for inclusion on the UNESCO World Heritage List as one of the most graphic examples on Earth of the geodiversity of dispersed volcanic fields in intracontinental settings. The book will culminate in a discussion of the geoheritage values of the volcanic fields of Saudi Arabia, providing some future research directions and development suggestions that could be the basis of sustainable geotourism in the region (Fig. 1.3).

1.1 Definition of Geosite

A hierarchical system has been established to arrange geological features for geoconservation, geoeducation and geotouristic purposes. The smallest unit in this system is the geosite (El Wartiti et al. 2008; Fuertes-Gutierrez and Fernandez-Martinez 2012; Gogin and Vdovets 2014; Joyce 2010; Kazanci 2012; Lansigu et al. 2014) or geomorphosite (Ahmadi et al. 2013; Bollati et al. 2013; Comanescu 2010; Coratza et al. 2011; Erhartic 2010; Feuillet and Sourp 2011; Harmon and Viles 2013; Panizza 2001; Pelfini and Bollati 2014; Pellitero et al. 2011; Pereira and Pereira 2010), depending on whether it is the geological or the geomorphological aspect of the identified site that is emphasized. Geosites should be single geologically and/or geomorphologically

Fig. 1.3 Small volcanoes pitted the western margin of the Arabian Peninsula such as the region nearby Al Birk town along the SW corner of the Kingdom of Saudi Arabia (GoogleEarth satellite image). *Top right corner* coordinate is 18° 4′ 0.73″N; 41° 52′ 29.56″E

important features that not only deserve protection (Martinez-Torres et al. 2011; Pica et al. 2014; Pulido Fernandez et al. 2014; Tomic and Bozic 2014; Vujičić et al. 2011), but also are good candidates to be part of a geoeducational program and geotouristic activity (Dryjanska 2014; Eder 1999, 2008; Farsani et al. 2014a, b; Hurtado et al. 2014; Lim 2014; Miccadei et al. 2014; Olafsdottir and Dowling 2014; Wang et al. 2014). Geosites are the core features of any geoeducational, geoconservation and geoheritage programs developed at geoparks (Eder 2008). Geosites should be defined in the context (and scale) of the region they are found in (Brocx and Semeniuk 2007). For example, in some regions a given geological phenomenon might meet the requirements to be a geosite, while the same phenomenon elsewhere would not generate enough attention from experts (and the public) to be viewed as a geosite. For instance, a sand dune could be defined, protected and promoted as geosite in Central Europe, while the same geological (geomorphological) feature would never be defined and listed as a geosite in the Kingdom of Saudi Arabia. However, if the same sand dune in the Kingdom of Saudi Arabia were to have some specific features, or were globally significant (e.g. by its size, volume, movement, composition etc.), it could then be defined as a geosite. This example highlights the importance of the scale and assessment method applied when evaluating a geosite and listing it in various inventories at the local, regional, or international level (Brocx and Semeniuk 2007). Evaluation methods for geosites are diverse and there is no consensus among researchers on what method is best. Specific definitions for geosites are abundant but most of the definitions capture similar elements of the following definition (Ilies and Josan 2009):

> The geosites are landforms with a specific shape, which alone or in collaboration with other bioecological or anthropic elements can become objects of heritage.
>
> Scientifically, the geosites are the most clear representation of the geomorphological processes, of the existent relations between the numerous factors which lead to their occurrence.

The above definition is based on an earlier view of sites that are important from a geomorphological perspective; therefore, it focuses strongly on the morphological view-point, as reflected in the use of the terms landforms and geomorphological processes (Panizza 2001). It is obvious that geosites are complex features that have the morphological aspects of a landscape as in the above outlined definition, but they also have a genetic relationship to the processes that formed the landform. Geomorphosite and geosite are often used in an undifferentiated way, although the first constitutes a component of the other (Ieleniz 2009). It is an interesting that the process of defining, assessing and evaluating a geosite is often heavily dependent on the purpose for which the definition and categorization is needed (Ieleniz 2009). The "scale" therefore depends on two main aspects: (1) the final purpose of the activities that request the identification of the geosites (related to science, tourism, the economy etc.), and (2) on the person (people) who realize the assessment (Ieleniz 2009). This dual approach is taken in to consideration with the identification of volcanic geosites from Saudi Arabia. To date, the research on volcanic geosites in Saudi Arabia has been primarily conducted to evaluate their scientific value in the context of their global scientific significance. Ieleniz (2009) mentioned that a geosite inventory can be slightly modified by different scaling, where the listed scientific information could be taken further and used to identify key geotouristic or even sustainable economic benefit projects in the future. Also, the people who have, to date, addressed geoheritage questions in Saudi Arabia have strong volcanological and geological backgrounds; hence, the listing of geosites and their geoheritage values will be presented through this perspective. Therefore, a possible development in the future might be that in the same region new geosites—even volcanic geosites—could be listed differently, in accordance to the background of subsequent evaluators (Ieleniz 2009). However, a thorough evaluation that follows a standardized procedure, can still and should still be heavily based on the geological aspects of the region (Ieleniz 2009). It is evident from the literature that evaluation methods are diverse and complex (Costa 2011; Gavrila and Anghel 2013; Ieleniz 2009; Ilies and Josan 2009; Mansur and Carvalho 2011; Panizza 2001; Petrovic et al. 2013; Tomic and Bozic 2014; 박준형 and Cheong 2012).

A revised definition of a geosite could be as a location that can offer some geological features that can help to understand a geological process, regardless of whether the location has any significant landform aspects, and/or can help to comprehend a geohazard feature and its influence on the natural and human environment. The latter aspect is particularly important for volcanic geosites, as many such sites would provide some details of the genetic process behind a particular form of volcanism, but these are not necessarily landform building elements. In a very simple definition **geosites are, therefore, geological features with a specific origin, appearance, and geohistorical attribute, which alone, or in collaboration with other bioecological or anthropic elements, can become objects of geoheritage**. Geomorphosites could be defined in a similar way to put more emphasis on the feature's geomorphological significance; a definition such as this would follow the concept of Ieleniz (2009). Overall the distinction between geosites and geomorphosites is not of great importance for identifying and evaluating the geoheritage of a site, providing it is measured and assessed from global geological perspectives and specifically designed scientific purposes.

The Kingdom of Saudi Arabia is rich in young volcanoes that erupted in the past 10 million years, particularly along its western margin, which is literally pitted by volcanoes. Despite their abundance, the individual geological value of many of these volcanoes (including the landscape-forming aesthetic value) makes these features stand out as potential geosites, because their unique geology can further the understanding of the formation of small-volume intracontinental volcanoes that together form a volcanic field (i.e. a harrat). If these identified geosites provide a coherent set of geological elements, they could be grouped together to form a geotopes. Various geological (volcanic) precincts can be ordered in a logical set of geotopes containing numerous geosites; these precincts together provide the basis for a geopark.

1.2 Definition of Geotope

The definition of a geotope is quite complicated in the literature. Linguistically it means more or less the same as geosite as tope means site or place, which suggests that the word refers to a small geographic spot (or spots) on the Earth's surface with a specific geological/geomorphological feature (loosely after Ieleniz 2009). Ieleniz (2009), therefore, defines a geotope as an indivisible geographic unit with a specific makeup, structure and functionality, resulting from a genesis, evolution and combination of certain environmental elements, which are differentiated due to their significance. While this definition appears complicated, it does capture the main features of a geotope, referring it to a specific geographic unit and also suggesting that hierarchically it stands above a simple geosite. A somewhat similar definition can be read in Wikipedia (http://en.wikipedia.org/wiki/Geotope), where a geotope is defined as "*the geological component of the abiotic matrix present in an ecotope. Example geotopes might be: an exposed outcrop of rocks, an erratic boulder, a grotto or ravine, a cave, an old stone wall marking a property boundary, and so forth*". The geotope has also been used as an analogy to a biotope, but applied to a well-localized portion of the non-living environment.

Other definitions of geotopes also mention that they are complex geological features that represent numerous individual geosites that together define an interrelated and geologically diverse entity (Drandaki et al. 1997; Fassoulas et al. 2012; Ieleniz 2009). In volcanic regions, a volcano or volcanic cone can be defined as a geotope (Fig. 1.4), based on the fact that it is not just a single volcanic landscape element, but is also a geological site that hosts a great variety of volcanic rock units formed through a series of volcanic eruption episodes with multiple eruption phases, commonly separated by syn-eruptive erosional processes (Manville et al. 2009). A volcano, therefore, is a very specific geological element with specific processes behind its formation that can be linked together its boundary can be delineated

Fig. 1.4 A single volcano as a perfect geotope with numerous geosites such as their crater region, edifice section and foothill areas. Between these main volcanic regions deep gullies can expose the interior of the volcano providing access to see the sedimentary facies architecture of them. Coordinate of the volcano is 24° 10′ 52.96″N; 39° 52′ 58.84″E

through the lateral distribution of its associated eruptive products. In this sense a volcano can be viewed as a perfect analogy for a biotope and probably is among the most well-defined geological features that is ideally suited to the concept of a geotope. The various (rock) facies that form a volcano (with their distinct 3D distribution controlled by specific volcanic processes), therefore could be best defined as volcanic geosites, especially if they can be distinctively defined in space.

1.3 Definition of Precinct

Many specific geotopes that are naturally linked and/or are of the same type can be defined as a precinct. A precinct could be a well-defined group of geological features of a specific type that can be linked together through some common ground, such as their rock types, geological age, rock formations with specific linkages to cultural or biological elements or just to delineate a specific section of an area with multidimensional geoheritage values. There is no consensus or common ground for how geotopes are selected for a precinct, but the following examples can provide some hints for how some groups have formulated precincts in practical geoheritage, geoconservation and geotouristic programs. One of the well-functioning regions where the precinct concept has been applied is the Kanawinka Geopark in South Australia and Victoria (Joyce 2009, 2010). The Kanawinka Geopark has distinguished five precincts over its large area, each highlighting a specific geological concept (www.kanawinka.org.au): (1) Craters and Limestone Precinct, (2) Plateaus and Falls Precinct, (3) Coast and Caves Precinct, (4) Cones and Flows Precinct, and (5) Lakes and Craters Precinct. The definition of a precinct can include the concept of a visitor gaining experience of a specific geological concept, which could be through exploring the sections of a specific precinct following pre-designed geo-routes. Therefore, the precinct concept can be viewed as a practical way to transmit to visitors information on geosites to geotopes, as part of geotouristic and geoeducational programs. Naturally if an area's geoheritage values are too diverse, the geological precinct concept can help to define specific regions, grouping them through naturally developing geological features, as in the Kanawinka Geopark.

The precinct concept is potentially the ideal tool to apply in volcanic regions where the specific geological sites and their geotopes are far from each other and/or accessed easily versus by adventure tourism, such as the strato-volcanic flank and ring plain region versus the crater zones where adventure tourism is required to visit key sites. While this is a distinction made mainly for practical reasons, it also can be designed in geotouristic programs by following fundamental geological concepts. For example, in volcanic regions the proximal areas of a volcano can be separated from its distal sections and can demonstrate the geological concept of volcanic mass transportation and/or impact of volcanic eruptions on the landscape (and, therefore, could be included in volcanic hazard education programs).

The precinct concept is even more valid and useful for implementing geoconservation and geotourism in areas with dispersed volcanism. In dispersed volcanic regions—commonly referred to as volcanic fields (Connor and Conway 2000; Manville et al. 2009)—volcanoes are commonly spaced a few kilometres apart over a large region (Boyce 2013; Condit and Connor 1996; Németh and Martin 1999; Ulrich et al. 1989). In volcanic fields, it is common for specific types of volcanoes to cluster together, which lends itself naturally to designating these volcanoes as a volcanic precinct in a geoeducational and/or geotoursitic programs. Moreover, volcanic features can also form the basis of the geoeducational and geotouristic programs developed in a volcanic precinct. It is also a trend to develop volcano routes linking volcanic regions together following a basic geological concept (e.g. volcanic features, volcano types, eruption styles, common mineral heritage etc.), such as those proposed across Central Europe (Harangi 2014). The volcanic regions of Saudi Arabia are dispersed in space over hundreds of kilometres. Therefore, adopting the volcanic precinct concept to develop geoeducational programs is a sound strategy. In addition the diversity of volcanic phenomena and their spatial distribution also justifies the development of volcanic precincts in the geoheritage sites of the Saudi Arabian volcanic regions.

1.4 Definition of Geopark

A geopark can be viewed as an umbrella under which defined geosites, geotops and precincts can be organised and their geological values conserved and offered to the public through geoeducational programs. A geopark in a volcanic region, therefore, could be organised along a common concept, such as a specific type of volcanism and highlighting its link to the human society that has been affected by the volcanism (Erfurt-Cooper 2014; Tefogoum et al. 2014; Wang et al. 2014; Woo et al. 2013). Geoparks may have a significant role in the economic development of rural areas or a local economy through geotourism (Dowling 2011; Farsani et al. 2012; Farsani et al. 2014a, b). They are also complex geoeducational sites, where fundamental knowledge can be passed to the general public on how the Earth works (Eder 1999, 2008; Keever and Zouros 2005; Mazumdar 2007; Zouros 2004; Azman et al. 2010, 2011).

The separation and definition of the geoconservation-geoeducation- and geotourism-defined units seem simple, but there are numerous obstacles and debates, especially how specific geological and geomorphological features should be selected as geosites/geomorphosites or how they could be identified as geotopes and eventually how they should be grouped to define a geopark (Brilha 2015; Brilha et al. 2005; Deraman et al. 2009, 2010; Fang and Tang 2010; Giusti and Calvet 2010; Joyce 2010; Liu et al. 2010; Strba et al. 2015). The distinction between geosites and geotopes is debatable and/or their definitions evolve in time as a reflection of the progress of the geoconservation of the region and the evolution of the scientific research. This means that the evaluation and specific grouping of volcanic features might change over time and different volcanic precincts may evolve in the future as a reflection of changing significance with the evolution of Earth Science. A particularly defined volcanic precinct could be expanded, decreased or redefined in the future, and therefore the later book chapters should be viewed as a guide to see how such a geoeducation program could be organized into a system that is scientifically well-supported and reflects the present day's concept of volcanism and understanding of the origin of the Saudi Arabian volcanism.

1.5 Definition of Geoheritage

The term geoheritage derives from the word heritage, which means something that has been transmitted from the past, or has been handed down by oral or written traditions (Brocx and Semeniuk 2007). The term also carries a notion of the heritage of features of a geological nature. It axiomatically conveys the ideal that there is something (valuable or otherwise) to inherit from the past and pass on to the future (Brocx and Semeniuk 2007). The best definition of geoheritage (Brocx and Semeniuk 2007) can be expressed as

> Globally, nationally, state-wide, to local features of geology, such as its igneous, metamorphic, sedimentary, structural, geochemical, mineralogical, palaeontological, geomorphic, pedologic, and hydrologic attributes, at all scales, that are intrinsically important sites, or culturally important sites, that offer information or insights into the formation or evolution of the Earth, or into the history of science, or that can be used for research, teaching, or reference.

Geoheritage focuses on the diversity of minerals, rocks and fossils, and petrogenetic features that indicate the origin and/or alteration of minerals, rocks and fossils. It also includes landforms and other geomorphological features that illustrate the effects of present and past climate and Earth forces (Brocx and Semeniuk 2007). Geoconservation derives from geoheritage, in that it deals with the conservation of Earth Science features, and therefore the evaluation of geoheritage directly feeds conceptual models to design geoconservation programs. Globally, geoheritage has become important because it is recognised that the Earth's systems have a story to tell, and that they are linked to the ongoing history of human development, providing the resources for development, and a sense of place, with historical, cultural, aesthetic, and religious values (Brocx and Semeniuk 2007).

In recent decades, there has been a global drive to preserve the heritage of the Earth (which we term here "*intrinsically significant sites of geoheritage*"), and to preserve the history of science embodied in some classic locations (which is termed as "*culturally significant sites of geoheritage*"). This conceptual framework provides the base to separate main approaches to identify geoheritage values.

Initially geoheritage sites were exclusively assigned and measured by the outstanding scientific importance or pure aesthetic beauty, and only recently have new scaling methods emerged to develop a more objective and gradated system to evaluate a geological feature's geoheritage value. Conservation sites of geoheritage significance are frequently found in regions where geological processes have created exceptional aesthetic or tourism values, or where the geology has created vegetation cover with high biological importance. From this perspective, volcanic regions are good candidates to have geoheritage values with high significance, as they equally can show how historic eruptions affected the human population to various degrees, influenced our agriculture and can help understand the complex volcano-petrologic processes that form various type of igneous rocks and associated pyroclastic successions. In addition, volcanoes can develop in a great diversity of geotectonic and sedimentary settings, providing a unique window in to the interaction between various geological processes. These facts make volcanoes attractive geological features that can capture the imagination of many, and can be primary tourist destinations, as stated by Erfurt-Cooper (2014).

The past two decades have seen a global effort in numerous countries to develop an inventory where every geological feature is listed and evaluated using various methods to express their geoheritage significance on various scales (Comanescu 2010; Deraman et al. 2010; Fang and Tang 2010). Geoheritage values traditionally evolved in different pathways in different regions, following the major trend of the development of Earth Science in the region and the utilization of the resources the Earth can provide. In areas such as Europe, where the industrial revolution occurred hundreds of years ago, and overpopulation and reduction of natural habitats became a burning issue, geoheritage studies attempted to classify and list in various inventories the geoheritage significance of almost every geologically interesting location, with an aim for such sites to be excluded from further destruction (Brocx and Semeniuk 2007). Behind the development of such inventories, there is always a geoconservation motivation (Brilha et al. 2005). In contrast in regions where the exploitation of natural resources is still at its maximum, geoheritage studies are driven by resource-exploitative ideas, and the progression of such inventories is strongly dependent on natural resource exploitation processes, such as is the case in Australia (Brocx and Semeniuk 2007).

In general, the evaluation of geoheritage values is arranged around the key aspects of the ***scope***, ***scale*** and ***significance*** of the identified geological feature (Brocx and Semeniuk 2007). One of the most important aspects of geoheritage and geoconservation is the identification of the scope of the geology. Geoheritage and geoconservation are concerned with geology and therefore the evaluation of geoheritage values should be based on the identification of the scope of the geological scientific attributes (Brocx and Semeniuk 2007). For volcanic geology, this would mean that the main driving force of the identification and grading of the geoheritage values of volcanic regions should be driven by the level of in-depth information provided by a volcanic geosite in order to understand the causes of the volcanism that created the identified site. Therefore, the evaluation of the significance of a site would be strongly driven by the level of available knowledge on the identified geosites. In the case of lack of sites that lack scientific research, because they are out of the mainstream research regions, the evaluation process would be strongly dependent on the level of scientific understanding and experience of the evaluator. This is a situation commonly faced when identifying the geoheritage values of the volcanic regions of western Saudi Arabia. The scientific research on many sites was limited, and the evaluation of the sites' volcanic geoheritage values was strongly controlled by the level of knowledge of similar volcanism by the group of researchers who carried out the evaluation.

Geology as science fundamentally follows two main directions: one focuses on the *geological history of the Earth* (time-dependent), and the other is *strictly associated with casual processes* (time-independent) (Fig. 1.5). This means that anything on the Earth's surface could either be linked to the way that feature formed through the Earth's history or provide a unique example to processes that take place even today (ie. are time-independent). When identifying volcanic geoheritage, these two approaches can be illustrated well. Time-dependent features can be any volcanically significant geological remains that can be linked to a specific volcanic

Fig. 1.5 Time-dependent (**a**) and time-independent (**b**) geoheritage values. Time-dependent values commonly associated with known volcanic eruptions such as the 1256 AD eruption site (**a**) [24° 21′ 31.86″N; 39° 46′ 13.52″E] while time-independent values are commonly related to stratigraphy or volcanic process-oriented sections such as the tuff ring succession near Harrat Hutaymah maar crater (**b**) [26° 59′ 41.14″N; 42° 14′ 46.90″E]

Fig. 1.6 1256 AD volcanic eruption site near the city of Al Madinah on GoogleEarth image. Note the young surface texture of the lava flow field just east from Al Madinah city and the well-preserved chain of cones in the *right bottom corner* of the image

eruption (e.g. commonly to a historic eruption that had some influence on human society and cultural development), while time-independent features could be any geological form that can tell a significant story on the formation of that feature, which could occur any time in the future if the conditions are favourable (Fig. 1.5). There are occasions when the two approaches meet in a single geotope and its geosites; this is the case with the 1256 AD volcanic eruption site in Saudi Arabia (Fig. 1.6). This site provides a unique window to the latest and historically well-documented eruption that resonates through many written documents, but also provides a spectacular site where the results of basic volcanic processes can be observed; these include lava fountaining, lava flow formations and widespread basaltic ash eruptions.

Geoconservation involves the conservation of sites with geological significance, but also deals with matters of environmental management, geohazards, sustainability, and natural heritage as it relates to maintaining habitats, biodiversity, and ecosystems in general (Brocx and Semeniuk 2007). Geodiversity, as understood by geoscientists, defines the

Fig. 1.7 Low geodiversity of monotonous lava fields of the harrats of Saudi Arabia still carry high geoheritage significance as the vast volume of the flow fields can provide an exciting and scientifically important story for understanding the formation of intra-continental volcanic fields [24° 23′ 20.78″N; 39° 46′ 4.31″E]

diversity of geological features on the Earth's surface, and connotes the variety of features within geology (Brocx and Semeniuk 2007). A different definition is sometimes used, which considers geodiversity to be a site-specific or region-specific term denoting the natural variety of geological, geomorphological, pedological and hydrological features of a given area. Combing these definitions gives a broad definition of geodiversity (Brocx and Semeniuk 2007) as *the natural variety of geological, geomorphological, pedological, hydrological features of a given area, from the purely static features at one extreme, to the assemblage of products, and at the other, their formative processes*. It is also important to emphasize that, while biodiversity can be viewed as a signature of the health of an ecosystem, the same cannot be applied to geodiversity. Low geodiversity is not a parallel to low biodiversity and does not carry the same negative connotations. Low geological diversity can be as equally important as high geological diversity and can have as good a case to be ranked as a geoheritage site with high significance. For instance, in a sedimentary geology context, a succession of a great thickness of otherwise "*not too exciting*", monotonous turbidity sand-silt-and-mudstones in a flysh belt (Alexandrowicz and Margielewski 2010; Ballance 1974) probably has low geodiversity. However, if such deposits accumulated over millions of years, its geological setting can tell a fantastic story and the site can be assigned a high geoheritage value in specified scales. In a volcanic context, such as the western Saudi Arabian Cenozoic volcanic region, the vast amount of lava flows can form a region with relatively low geodiversity, composed of dark monotonous lava fields (Fig. 1.7). This low geodiversity does not mean the region has a low geoheritage value. On the contrary, the total volume of magma that erupted and formed the lava flow fields provides a conceptual model for the formation of the Saudi harrats, and therefore such lava fields can be graded as geoheritage sites with high significance.

In a geotectonic sense, a complicated geological setting (e.g. collision zones of continents) have often scored highly in estimates of geodiversity, on the basis of the diverse nature of the geological features that can be identified in such regions. Consequently such regions are often considered to have the higher geoheritage values than any other relatively passive geological region (e.g. intracontinental basins). This is not a sound approach and should be avoided.

When evaluating geoheritage significance, it is important to define the scale of observation and evaluation. These can be *regional*, *large*, *medium*, *small*, *fine* and *very fine*. The identification of the scale is important; a region can have a very high geoheritage significance at a microscopic scale (e.g. a rare mineral associated with a volcanic deposit), while on a medium, large or regional scale the same site may be insignificant. In general the scale could be best defined as a mountain range being regional scale, while the very fine scale could be a single crystal or fossil.

For evaluating the geoheritage values of a geological feature in a specific scale, its significance must be identified (Brocx and Semeniuk 2007). This is a difficult task. The level of importance normally attributed to a given feature of geoheritage significance is related to one of two factors: (1) ***how frequent, or common, the feature is within a scale of reference***, and (2) ***how important the feature is intrinsically or culturally***. Putting this system in a volcanic context, the first point refers to how common the volcanic feature is. For instance, a scoria cone is a very common feature in every intra-continental geological setting and is described as the most common volcanic landform not only on the Earth but also in the entire Solar System (Németh et al. 2011). Hence, a scoria cone would be downgraded in this aspect. However, if we look at the number of scoria cones on a regional scale, a volcanic field, where there may be hundreds of scoria cones that erupted over a relatively short period of time, the same volcanic feature (in their groups) would be graded as a geologically rare feature and the scoria cone field would score highly in the significance table on the regional scale for the second factor above. Furthermore, a scoria cone, which has specific beds abundant in a specific size of xenoliths carried great variety of fragments from the upper mantle, would also be defined as being a volcanic feature that is reasonable rare and thus would score highly in terms of its intrinsic importance. This highlights the relative, and probably time-dependent, nature of the evaluation of any geological feature and it should be viewed as a value that can change over time and scale. The evaluation is also strongly dependent on the evaluators' background knowledge and research intensity and thus ability to explore and measure the geoheritage value against the best possible broad scale and consider the latest developments of the specific geoscience fields in which studied geoheritage sites fall.

Five levels of significance are used: (1) ***International***, (2) ***National***, (3) ***Large regional***, (4) ***Small regional*** and (5) ***Local***. These significance levels can be defined in various ways. For example, (Brocx and Semeniuk 2007) state that something has international significance if it is "*only one, or a few, or the best example of a given feature occurring globally, hence it is globally unique, rare, or uncommon; or performs a function in a global network*". Brocx and Semeniuk (2007) also define something with local significance as "*the natural history feature is important only to the local community*". Anything between these two end-members has a connotation to the frequency of the geological feature and to the scale of the observation.

1.6 Dissemination of Geoheritage Information and Geoeducation Strategy with Special Relevance to Saudi Arabia

1.6.1 Web-Based Information Sites

The most effective method to disseminate information today is to utilize internet-based resources, which have been widely used in many recent geoeducational, geoconservation and geotourism projects (Yan et al. 2008, 2009, 2010). To promote the geoheritage value of the volcanic regions of Western Saudi Arabia, a new website, called ***arabiageoparks.com***, is planned to be launched soon. This website will act as a resource center for visitors and/or end users (e.g. tour operators) of the volcanic geoheritage sites, providing easy to understand information on the volcanic processes and the individual volcanic geotopes' geosites that form the well-defined volcanic precincts that will be the basis of the proposed volcanic geoparks in the region. The utilisation of electronic resources is cost-effective and easy to access, as well as serving as a hub that links to geographical sites elsewhere on Earth that are relevant to the volcanic features and/or geoeducational purposes of the geoconservation and geoheritage sites of Western Saudi Arabia. In addition, the website will provide links to further material for visitors or researchers wishing to access more in-depth information on the geology and volcanism of the geoheritage sites. Such websites can be maintained easily and easily updated to include information on the listed geosites that may emerge in the future.

1.6.2 Educational Leaflets

While educational leaflets are considered by some to be "*old-fashioned*", they remain effective tools to provide basic, accurate information on specific types of geological features at any geoheritage site in the world, including geoparks. Preparation of such leaflets is strongly encouraged; however, it is important that they contain up-to-date information and procedures should be put in place to ensure this, especially in areas where scientific research is active, such as in Western Saudi Arabia. Leaflets should contain basic maps, GPS-coordinates, colour images, and very short descriptions of the expected geological features that a visitor may encounter. Such leaflets have been successfully prepared for the Kanawinka Geopark in Australia and provide easy-to-access, map-based information that any independent visitor can use. These leaflets can also provide basic health and safety information, which is especially necessary for independent geotourists wishing to visit the remote areas of the volcanic geoheritage sites of Western Saudi Arabia.

1.6.3 Educational Boards at Entry Points to Sites and Main Regional Transportation Hubs

Educational boards are widely used in many geoparks and nature conservation sites worldwide and they are considered an effective tool for transferring knowledge to visitors (Kazanci 2012; Moreira 2012). This form of information dissemination is particularly effective for ad-hoc visitors. While such resources are desirable, they are generally costly, require regular maintenance, and they are difficult to update as new information becomes available. In addition educational boards are susceptible to damage due to weather or vandalism. While vandalism is rather rare in the Kingdom of Saudi Arabia, the weather is a major issue since Western Saudi Arabia is located in an area where the summer temperatures commonly approach 50 °C, and flash floods and strong sand storms and wind blasts are also not uncommon. In the first phase of the development, educational boards should be established only at the main entry points to major geoheritage sites, particularly at the sites expected to be most visited. A similar set of boards to those on-site could be placed at major transportation hubs, such as airports and bus stations, where visitors are expected to arrive to visit the region.

1.6.4 Smart Phone Applications

Smart phones (and web-capable tablets, notebooks) are widespread, and increasingly accessible for the general public. In the Kingdom of Saudi Arabia the mobile phone network is well-developed, and in many major centers currently being upgraded for 4G capability. Mobile internet hotspots and/or free or low service charge WiFi internet access are common in many tourist accommodation spots, as well as in specific parts of towns. Mobile network coverage is good, in spite of the large distances and low population densities outside of major urban areas. The territory of significant volcanic geoheritage sites is well covered by mobile networks, in spite of the general remoteness of many of those sites. Internet access at geoheritage sites allows the visitor to download maps, and other information about the volcanic geoheritage sites when they are there. On the basis of this developed technological background, it is a logical to put most of the information on the volcanic geoheritage sites at a geopark onto a mobile-compatible website (as discussed above). The information should be in a form that is downloadable to enable visitors to download the information to their mobile devices prior to their visit, should they not wish to access the internet while in the park. The development of smart phone applications is also suggested. Such applications could include interactive maps, geosite descriptions, and a volcanological dictionary that are closely linked with real-time weather forecasts, sunset/sunrise times and moon phase information, as well as information on what to do in an emergency. These smart phone applications could be accessed at a nominal low fee, and should be available through mainstream application stores.

1.6.5 Organization of Information Flow Through Local and Global Travel Agencies

One of the most important aspects of the success of the geohertiage projects recently proposed in Saudi Arabia is the link between information providers and end users. This can be coordinated and achieved by working closely with responsible governmental agencies. To increase the number of international visitors, licenced designated travel agencies need to be linked with the activities offered at the proposed geoparks. To facilitate this, regular training sessions and workshops, with field visits, could be held, where, beside the geological information specific to the areas, basic training could be given on volcanology, nature and specifically geoconservation. These workshops should be designed, organized and provided regularly by an expert educational institute preferably linked to a university, such as the King Abdulaziz University in Jeddah, as Jeddah is a gateway to the Western Saudi Arabian volcanic regions. It is anticipated that a geoeducational program for tour operators as described here will ensure a slow but steady increase in the numbers of participants in nature conservation and geoeducation programs.

1.6.6 Global Links

The geoheritage sites of Western Saudi Arabia have many complementary features to those found in many other volcanic geoheritage sites (many of them organised in regional or global geoparks) around the world and therefore there is a unique opportunity to link the Saudi sites (and the associated geoheritage, geoconservation and geoeducation programs) with other geoconservation sites in similar volcanic fields worldwide. Examples of possible partner regions where established geoparks (on various levels) already exists include the Kanawinka Global Geopark in Australia (www.kanawinkageopark.org.au), Jeju Island Geopark (www.geopark.jeju.go.kr), Nograd/Novohrad Geopark on the border of Hungary and Slovakia (www.nogradgeopark.eu), Bakony-Balaton Geopark in Hungary (www.bakony-balaton-geopark.eu), and Vulkaneifel Geopark in Germany (www.geopark-vulkaneifel.de). Not all features of each of these geoparks are relevant to each of the geoheritage sites and proposed volcanic geoparks of Western Saudi Arabia, but many of them share similarities with the style, extent, age, impact and cultural relevance of the volcanism on show in the majority of the volcanic geoheritage sites of Western Saudi Arabia.

The 1256 AD and 641 AD historic eruption sites near Al Madinah City, as part of the Harrat Al Madinah, and its proposed geoeducational programs, could be easily linked to other similar sites around the world, such as the Bakony-Balaton Geopark in Hungary, and the Wudalianchi Geopark in China (http://english.wdlc.com.cn). The chain of well-exposed volcanic cones in the Craters of the Moon National Monument in Idaho (www.nps.gov/crmo/) also shows similar volcanological features to the 1256 AD and 641 AD historic eruption sites. In addition, a similar volcanic eruption scenario to that inferred to have produced the 1256 AD and 641 AD eruption sites in Western Saudi Arabia is known from a few volcanic cones on natural protected areas in Lanzarote (Carracedo et al. 1992; Kervyn et al. 2012) and Tenerife (Kereszturi et al. 2012; Paez 2010), both in Canary Islands, Spain, or Laki Fissure in Iceland (Thordarson and Self 1993). Scoria cones and similar intracontinental monogenetic volcanic landforms are organized and arranged in geoparks in the Eifel, Germany (www.geopark-vulkaneifel.de) and in the Kanawinka Geopark in Australia (www.kanawinkageopark.org.au); however, these sites are generally covered with dense vegetation or grassland and, while the original volcanic landforms are young and well-preserved, the accessibility and visibility of the syn-eruptive volcanic landforms are more difficult than in the proposed volcanic geoheritage sites in Western Saudi Arabia. In addition, the Nograd/Novohrad Geopark on the border of Hungary and Slovakia (www.nogradgeopark.eu) has volcanic features associated with monogenetic volcanism, but these can only be seen on eroded volcanic landforms, where the core of a scoria cone, lava spatter cone or tuff ring core has been exhumed due to erosion. Similar connections to the volcanic geoheritage sites of Western Saudi Arabia can be drawn from the protected volcanic sites of the Auckland Volcanic Field in New Zealand or the UNESCO World Heritage Site of Jeju Island in Korea. However, these sites are also heavily vegetated and the syn-eruptive volcanic features are commonly obscured. Links between these sites and the Saudi volcanic geoheritage sites would allow awareness to be raised in visitors spending time exploring similar, but vegetation-covered, volcanoes in more humid climatic regions. Many of the Western Saudi Arabian volcanic geoheritage sites can be linked to protected volcanic regions, such as those in Hawaii or in the San Francisco Volcanic Field in Arizona, which are considered by the scientific community as the type localities to illustrate the consequences of lava fountaining, pit crater formation and lava spatter cone evolution. The Western Saudi Arabian volcanic regions can provide a great abundance of such volcanic landforms in exceptionally well preserved conditions. Other volcanic features, such as explosion craters, silicic lava domes and various silicic lava flows, in Western Saudi Arabia could be linked to geoheritage sites located in globally well-known sites, such as those in the Unzen Volcanic Area Global Geopark in Japan (www.unzen-geopark.jp). While the geological setting is strikingly different between the volcanic regions of Western Saudi Arabia and Unzen, the resulting volcanic landforms and associated volcanic eruption styles are very similar. A more suitable link could be promoted between the volcanic regions of Western Saudi Arabia and Jeju Island in South Korea, which hosts the Jeju Island Geopark and UNESCO World Heritage site (Sik et al. 2013; Sohnyoungkwan et al. 2009; Woo et al. 2013). Jeju Island is the home of numerous silicic lava domes and scoria cones that together make an extensive volcanic field that formed in the past 1 million years (Brenna et al. 2012).

1.7 Volcano Types and Volcanic Geoheritage Relevant to the Volcanic Regions of Saudi Arabia

The establishment of volcanic geoparks and geoheritage sites is becoming increasingly popular worldwide (Armiero et al. 2011; Bitschene and Schueller 2011; Erfurt-Cooper 2011; Joyce 2009). In addition, UNESCO Global Geoparks (www.globalgeopark.org) and the European Geopark Network (www.europeangeoparks.org) have numerous geoparks that have primarily achieved their status due to their volcanic geoheritage (e.g. www.geopark-vulkaneifel.de; www.bakony-balaton-geopark.hu; www.nogradgeopark.eu). Among these sites there are locations where the main attraction of the geopark is the large number of small-volume volcanic edifices in various geotectonic settings, often with strong interactions with human societies. Here we provide a summary of the volcano types, their eruption styles and significance with relevance to the identified volcanic geoheritage sites of Saudi Arabia.

Small-volume volcanoes are commonly defined as monogenetic, which is a reflection of (a) their short-lived eruption styles that are commonly associated with the arrival

Fig. 1.8 A typical landscape with small-volume volcano from Western Saudi Arabia [view taken from 24° 12′ 8.54″N; 39° 52′ 42.46″E toward SSE]

of a single magma batch to the surface and (b) their generally small volcanic edifices (edifice size ranges are normally between 0.001 and 0.1 km^3) (Kereszturi and Németh 2012; Németh 2010; Németh and Kereszturi 2015; Valentine and Gregg 2008; Walker 1993). The small size of these volcanoes makes them relatively easy to access; their eruptive products are on a "*human scale*" (Fig. 1.8) and therefore they can be used to demonstrate fundamental volcanic processes without major logistical challenges for education program designers or for visitors.

A common small-volume monogenetic volcano is scoria (cinder) cones (Fig. 1.9); these are constructional volcanic edifices and normally result from the mild explosive eruption of magma, which is triggered by the bubble coalescence of volatiles in the upper conduit of the cone at moderate magma discharge rates (Kereszturi and Németh 2012; Vergniolle 1996; Vergniolle and Brandeis 1996; Vergniolle and Manga 2000). At high magma discharge rates, when bubbles travel together with the magma, lava fountaining can occur, which can build up lava spatter cones (Vergniolle and Manga 2000). In scoria and lava spatter cone building events, the volcanic eruption is primarily governed by the so-called internal controlling parameters, which are characteristic of the magma itself (e.g. viscosity, volatile content, chemistry, discharge rate etc.) (Vergniolle and Manga 2000). The resulting volcanic edifices are constructional and form randomly or systematically distributed vents over a volcanic field (Connor 1990; Connor and Conway 2000). Their sizes, edifice structures and deposit characteristics are primarily linked to the magma internal parameters and the structural elements of the lithosphere the magma encounters en-route to the surface (Fig. 1.9).

Near the surface (<2 km), the basement rocks are fractured and commonly filled with water-forming hydrologically active zones and/or sedimentary basins that are filled with unconsolidated water saturated clastic material. If the hot magma reaches and encounters these water-rich regions, it can interact at various levels with the water hosted in the country rock and deposits, leading to phreatomagmatic explosive eruptions (Lorenz 1986; White and Ross 2011). Alternatively magma can reach the surface, where shallow sub-surface aquifers charged with water and/or surface water bodies (and/or water-saturated sediments) are abundant (White and Ross 2011). The magma can interact with the water, producing phreatomagmatic explosions that can form strikingly different deposits and volcanic landforms to those formed purely by the internal parameters controlling the explosivity of the rising magma (Fig. 1.10) (White and Ross 2011).

Phreatomagmatism has been widely recognized across many volcanic fields, even among those dominated by

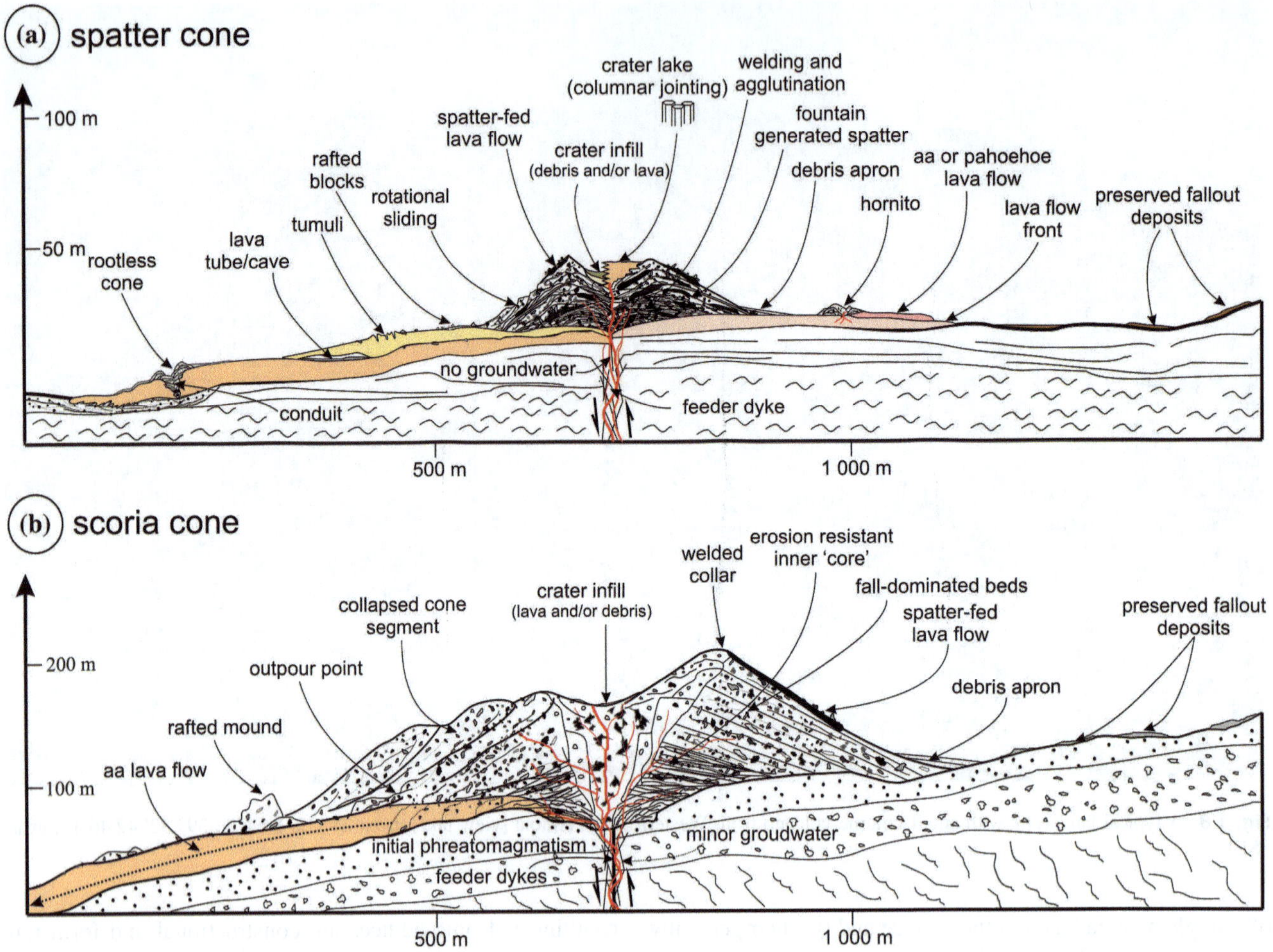

Fig. 1.9 Monogenetic volcanic landforms associated with "dry" eruptions and their basic features relevant to small-volume volcanoes of Saudi Arabia after Kereszturi and Németh (2012)

volcanic landforms typical of explosive eruptions triggered by the magma's internal physico-chemical conditions (Risso et al. 2008). Typical phreatomagmatic landforms are maars and tuff rings (Fig. 1.10) in intracontinental settings (Kereszturi and Németh 2012). Maars are the result of explosive fragmentation of magma due to contact with ground-water, while tuff rings are more typical of eruptions where magma interacts with shallow sub-surface water and/or surface standing water bodies (Vespermann and Schmincke 2000). The resulting volcanic landforms are different from the constructional landforms of scoria (cinder) cones. Maars have deep craters and potential diatremes beneath the crater (the crater cuts deep into the syn-eruptive surface and therefore pre-eruptive lithologies are exposed in the crater wall), while tuff rings are broad volcanic craters with low crater rims that normally sit on the syn-eruptive landscape (White and Ross 2011). While phreatomagmatism is the largely accepted mechanism for the formation of maars and tuff rings (White and Ross 2011), there are ambiguous cases where maars and tuff rings may have been formed due to an extreme volatile content of alkaline-rich melts (Stoppa and Schiazza 2013). In basaltic systems, however, there is a wealth of field evidence to demonstrate that phreatomagmatism is the main force generating these volcanoes (White and Ross 2011). Therefore, recognizing maars and tuff rings in a volcanic field can have significant implications in terms of understanding the hydrology and environmental (external) conditions at the time of volcanism (Agustin-Flores et al. 2014; Kereszturi et al. 2011, 2014; Németh et al. 2012). The balance between the external and internal forces and the resulting differing eruption styles, producing different eruptive units, is particularly important in the evaluation of the geoheritage values of volcanic geosites in Western Saudi Arabia. The potential geoheritage sites in this region could be the centre of global research effort to understand link between volcanic edifices and the potential eruption styles that formed them.

In addition, the general longevity of many of the volcanic fields in Saudi Arabia (in the range of thousands to millions of years) means that the volcanic regions have likely

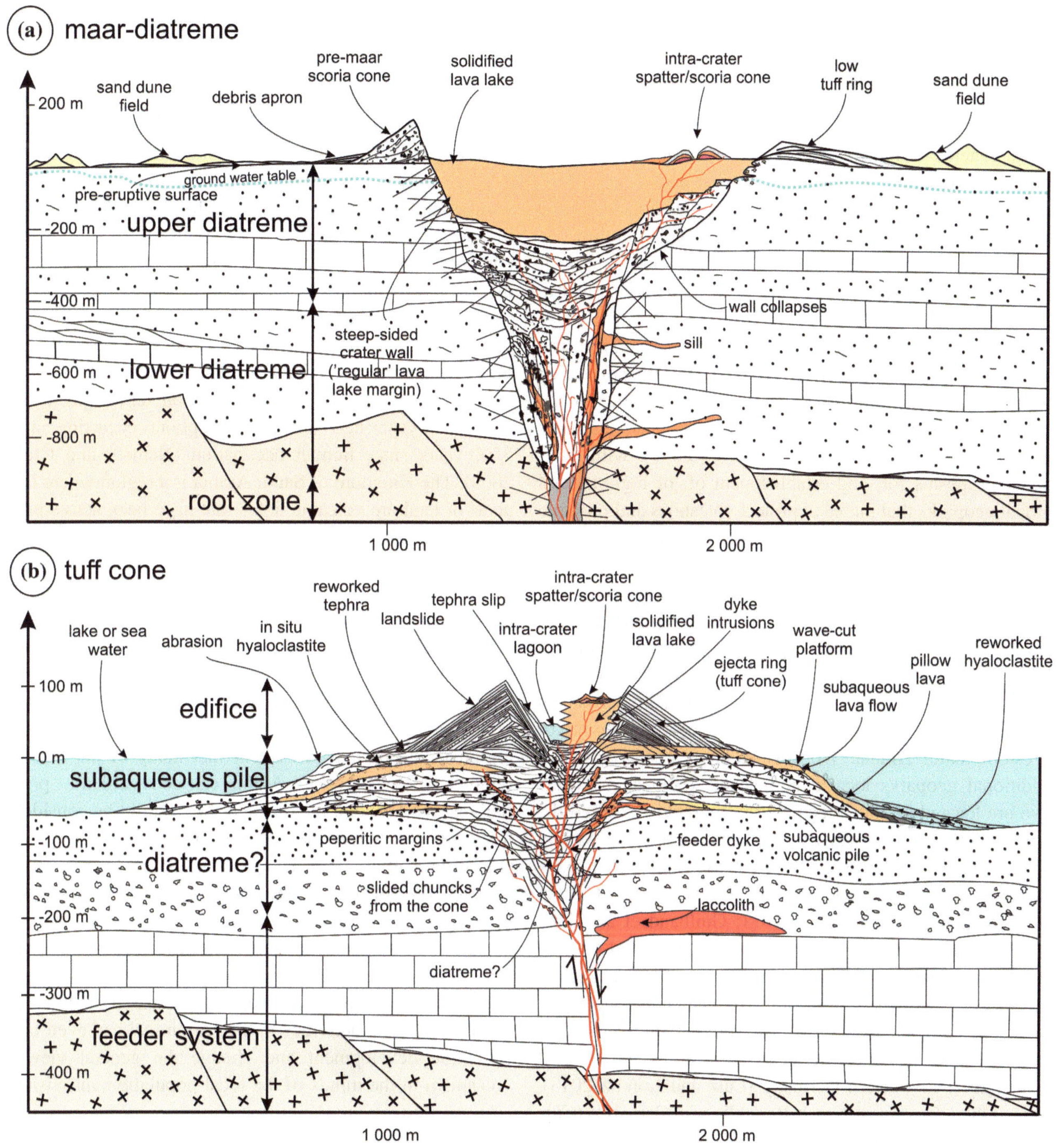

Fig. 1.10 Monogenetic volcanic landforms associated with "wet" eruptions and their basic features relevant to small-volume volcanoes of Saudi Arabia after Kereszturi and Németh (2012)

experienced environmental changes. The period when volcanism was commonly phreatomagmatic, for example, could be interpreted as a more humid climatic period to the otherwise dry and arid conditions of today. Thus such geoheritage sites in Saudi Arabia could be used for very important geoeducation initiatives to demonstrate to the visitors that the climate can change dramatically over a long period of time (Moufti et al. 2014). There are several examples worldwide where there are volcanic landforms that can be connected to a period in the evolution of the region when more humid climatic conditions were common (Kereszturi et al. 2011). While the overall type and structure of a monogenetic volcanic edifice is the result of an extremely complex process, the general abundance of phreatomagmatic volcanic edifices in a volcanic field is indicative of wetter, and/or a hydrogeologically more active, period. Volcanic geosites where such eruption style variations can be detected could be key volcanic geoheritage sites in Saudi Arabia and provide excellent geoeducational material.

In general, the increase in research targeted at understanding small-volume volcanoes globally and the wealth of new knowledge on such volcanism can be directly channelled to the general public through geoeducational projects. In this respect Saudi Arabia has an exceptional supply of globally significant geosites, which will be demonstrated in this book.

The developing trend in volcanic geoheritage research and associated geoeducational programs through protected volcanic geosites and landscapes is clearly visible in the increasing number of studies on volcanic geosites and geomorphosites (Erfurt-Cooper 2014; Joyce 2009), with additional proposals for, and establishment of, of regional and global geoparks that intend to protect volcanoes and promote our understanding of monogenetic volcanism (Armiero et al. 2011; Joyce 2010). Currently the European Geopark Network has 52 official members from 18 countries (www.europeangeopark.org). Of these 52, at least five have a strong emphasis on volcanic geology, including geoparks located in Germany, Hungary, Iceland, Slovakia and Spain. Of these volcanic-oriented geoparks, three include volcanic features similar to those that form the volcanic regions of Western Saudi Arabia. Worldwide, there are at least five additional geoparks located in volcanic regions, of which two are located on monogenetic volcanic fields. Around the world, there are many heavily populated regions that occupy areas of active volcanic fields, such as Auckland in New Zealand (Bebbington and Cronin 2011), Mexico City in Mexico (Agustin-Flores et al. 2011; Guilbaud et al. 2009, 2012), or the Bay of Naples in Italy (Orsi et al. 2004). Relatively young volcanic fields are also located in culturally significant areas, where some of the fundamental concepts of monogenetic volcanism were discovered, such as those in Central France at the Chaîne des Puys (Camus and Vincent 1973), Eifel in western Germany (Lutz and Lorenz 2013), Eger (Ohre) Graben in Czech Republic and Germany (Cajz et al. 2009) and in the Pannonian Basin, mostly in Hungary (Martin and Németh 2004). With the growing population on Earth, it is becoming increasingly important to develop geoeducational programs in these regions and others to disseminate our current understanding of this type of volcanism, with an aim to pass on information about potential eruption scenarios, volcanic hazards, and available volcanic crisis management to the general public. Similar educational projects, in conjunction with geoconservation strategies, have been developed around other natural geohazards, such as faulting (Zouros et al. 2011), and a growing number of methodological studies have been reported that combine geoeducation with effective geoconservation, an objective that underpins geopark projects (Coratza and De Waele 2012). In volcanic areas that have not experienced volcanic eruptions in the recent past (e.g. no living memories, oral traditions or written documentation are available), such as many regions in Western Saudi Arabia, the usage of geoheritage sites is especially important to establish a link between the preserved volcanic landforms and their potential link to specific volcanic hazards. Such projects are also of particular importance in areas where the societal structure of the local communities is such that memories of recent volcanic eruptions can quickly vanish. An example of this is the young volcanic island of Western Samoa, where there is limited information available on the volcanic origins of the landscape, in spite of the last eruption taking place only about 100 years ago and many eruptions occurring in the 3500 years since human occupation (Németh and Cronin 2009). The Kingdom of Saudi Arabia is a region where large areas of land are volcanic fields that have been active in the past 10 million years (Moufti et al. 2013a, b), leaving behind nearly every volcanic feature possible in a dispersed intra-continental volcanic field. These regions have not experienced a volcanic eruption since the last event in 1256 AD (Camp et al. 1987). However, there is evidence, based on the morphologies of many of the volcanoes, that indicates that other historic eruptions may have occurred, but have gone unnoticed because of the low population density of the region in the past. It is evident that many of the volcanic regions in Western Saudi Arabia can be regarded as potentially active volcanic regions and they should be considered to be regions where the volcanic hazard is not negligible (El Difrawy et al. 2013; Runge et al. 2014). In this respect, research on volcanic geoheritage and the development of geoeducational programs through various activities and the involvement of geotouristic organisations can serve a vital role in volcanic hazard education in the region. Therefore, this should be viewed as an important research field that can form the basis of activities in the region that can even bring economic development and change the general view of volcanism in the minds of the local population and visitors.

References

Agustin-Flores J, Siebe C, Guilbaud M-N (2011) Geology and geochemistry of Pelagatos, Cerro del Agua, and Dos Cerros monogenetic volcanoes in the Sierra Chichinautzin volcanic field, south of Mexico City. J Volcanol Geoth Res 201(1–4):143–162

Agustin-Flores J, Nemeth K, Cronin SJ, Lindsay JM, Kereszturi G, Brand BD, Smith IEM (2014) Phreatomagmatic eruptions through unconsolidated coastal plain sequences, Maungataketake, Auckland volcanic field (New Zealand). J Volcanol Geoth Res 276:46–63

Ahluwalia AD (2006) Indian geoheritage, geodiversity: Geosites and geoparks. Curr Sci 91(10):1307

Ahmadi A, Esfehani M, Mikaeili P, Maghsodi A, Shorestan AG, Shokrizare H (2013) Geomorphosite assessment of Ziarat village by the use of scientific and additional values methods. J Am Sci 9 (3s):27–31

Alexandrowicz Z, Margielewski W (2010) Impact of mass movements on geo- and biodiversity in the Polish Outer (Flysch) Carpathians. Geomorphology 123(3–4):290–304

Armiero V, Petrosino P, Lirer L, Alberico I (2011) The GeoCaF project: proposal of a geosites network at Campi Flegrei (southern Italy). Geoheritage 3(3):195–219

Avram M, Zarrilli L (2012) The Italian model of "albergo diffuso": a possible way to preserve the traditional heritage and to encourage the sustainable development of the Apuseni Nature Park. GeoJournal Tourism Geosites 9(1):32–42

Azman N, Halim SA, Liu OP, Saidin S, Komoo I (2010) Public education in heritage conservation for geopark community. In: Jelas ZM, Salleh A, Azman N (eds) International Conference on Learner Diversity 2010. pp 504–511

Azman N, Halim SA, Liu OP, Komoo I (2011) The langkawi global geopark: local community's perspectives on public education. Int J Herit Stud 17(3):261–279

Ballance PF (1974) An inter-arc flysch basin in northern New Zealand: Waitemata group (upper Oligocene to lower Miocene). J Geol 82:439–471

Bebbington MS, Cronin SJ (2011) Spatio-temporal hazard estimation in the Auckland volcanic field, New Zealand, with a new event-order model. Bull Volcanol 73(1):55–72

Bitschene P, Schueller A (2011) Geo-education and geopark implementation in the Vulkaneifel European geopark. GSA Field Guide 22:29–34

Bollati I, Smiraglia C, Pelfini M (2013) Assessment and selection of geomorphosites and trails in the Miage Glacier Area (western Italian Alps). Environ Manag 51(4):951–967

Boyce J (2013) The Newer Volcanics Province of southeastern Australia: a new classification scheme and distribution map for eruption centres. Aust J Earth Sci 60(4):449–462

Brenna M, Cronin SJ, Smith IEM, Sohn YK, Maas R (2012) Spatio-temporal evolution of a dispersed magmatic system and its implications for volcano growth, Jeju Island volcanic field, Korea. Lithos 148:337–352

Brilha J (2015) Inventory and quantitative assesment of geosites and geodiversity sites: a review. Geoheritage. doi:10.1007/s12371-014-0139-3

Brilha J, Andrade C, Azeredo A, Barriga F, Cachao M, Couto H, Cunha PP, Crispim JA, Dantas P, Duarte LV, Freitas MC, Granja HM, Henriques MH, Henriques R, Lopes L, Madeira J, Matos JMX, Noronha F, Pais J, Picarra J, Ramalho MM, Relvas J, Ribeiro A, Santos A, Santos VF, Terrinha P (2005) Definition of the Portuguese frameworks with international relevance as an input for the European geological heritage characterisation. Episodes 28(3): 177–186

Brocx M, Semeniuk V (2007) Geoheritage and geoconservation—history, definition, scope and scale. J Roy Soc W Aust 90(Part 2): 53–87

Bruno DE, Crowley BE, Gutak JM, Moroni A, Nazarenko OV, Oheim KB, Ruban DA, Tiess G, Zorina SO (2014) Paleogeography as geological heritage: developing geosite classification. Earth Sci Rev 138:300–312

Cajz V, Rapprich V, Erban V, Pecskay Z, Radon M (2009) Late Miocene volcanic activity in the Ceske stredohori Mountains (Ohre/Eger Graben, northern Bohemia). Geol Carpath 60(6):519–533

Camp VE, Hooper PR, Roobol MJ, White DL (1987) The Madinah eruption, Saudi Arabia: Magma mixing and simultaneous extrusion of three basaltic chemical types. Bull Volcanol 49(2):489–508

Camus G, Vincent PM (1973) Le Chuquet Genestoux et les projections palagonitiques du maar d'Enval. Revue des Sciences Naturelles d'Auvergne 39(1–4):1–12

Cardoso CS, Batista SG (2013) Innovation of tourist offer based on local values: a study of the Geopark Serido, RN. Brazil Caderno Virtual de Turismo 13(2):150–161

Carracedo JC, Rodriguez Badiola E, Soler V (1992) The 1730–1736 eruption of Lanzarote, Canary slands: a long, high-magnitude basaltic fissure eruption. J Volcanol Geoth Res 53:239–250

Comanescu L (2010) The evaliuation of the geomorphosites' value. case study the Bucegi Mountains. Metal Int 15:212–216

Comanescu L, Nedelea A (2012) The assessment of geodiversity—a premise for declaring the geopark Buzului County (Romania). J Earth Syst Sci 121(6):1493–1500

Condit CD, Connor CB (1996) Recurrence rates of volcanism in basaltic volcanic fields: an example from the Springerville volcanic field, Arizona (vol 108, p 1225, 1996). Geological Society of America Bulletin 108(11):1514–1514

Connor CB (1990) Cinder cone clustering in the TransMexican volcanic belt: implications for structural and petrologic models. J Geophys Res 95(B12):19395–19405

Connor CB, Conway FM (2000) Basaltic volcanic fields. In: Sigurdsson H (ed) Encyclopedia of volcanoes. Academic Press, San Diego, pp 331–343

Coratza P, De Waele J (2012) Geomorphosites and natural hazards: teaching the importance of geomorphology in society. Geoheritage 4(3):195–203

Coratza P, Bruschi VM, Piacentini D, Saliba D, Soldati M (2011) Recognition and assessment of geomorphosites in Malta at the Il-Majjistral nature and history park. Geoheritage 3(3):175–185

Corvea JL, Blanco A, de Bustamante I, Farfan H, Martinez Y, Novo R, Diaz C, Lopez N (2014) Advances in geoconservation in Cuba: assessment of the Guaniguanico range and Guanahacabibes Plain (Pinar del Rio). Geoheritage 6(1):1–16

Costa FL (2011) Volcanic geomorphosites assessment of the last eruption, on April to May 1995, within the natural park of Fogo Island, Cape Verde. GeoJournal Tourism Geosites 8(2):167–177

Deraman A, Yahaya J, Salim J, Idris S, Jambari DI, Jaradat A, Komoo I, Leman MS, Unjah T, Sarman M, Sian LC (2009) The development of mygeo-rs: a knowledge management system of geodiversity data for tourism industries. pp 658–662. 978-0-9821489-0-7

Deraman A, Yahaya JH, Salim J, Idris S, Komoo I, Leman MS, Unjah T (2010) A geodiversity repository system based on digital object modeling: an applied research on Langkawi geoparks. In: Soliman KS (ed) Knowledge management and innovation: a business competitive edge perspective, Vols 1–3. pp 1208–1217

Doniz-Paez J, Becerra-Ramirez R, Gonzalez-Cardenas E, Guillen-Martin C, Escobar-Lahoz E (2011) Geomorphosites and geotourism in volcanic landscape: the example of La Corona del Lajial cinder cone (El Hierro, Canary Islands, Spain). GeoJournal Tourism Geosites 8(2):185–197

Dowling RK (2011) Geotourism's global growth. Geoheritage 3(1): 1–13

Drandaki IT, Koutsouveli A, Ioakim C (1997) Geological geomorphological heritage—geotopes. In: Marinos PG, Koukis GC, Tsiambaos GC, Stournaras GC (eds) Engineering geology and the environment Vol 1–3. Athens, Greece, pp 3021–3026

Dryjanska L (2014) Geotourism potential of the islands of the western Mediterranean: case study of Elba Island. Eur J Tourism Res 7: 127–139

Eder W (1999) "UNESCO GEOPARKS"—a new initiative for protection and sustainable development of the earth's heritage. Neues Jahrbuch für Geologie und Paläontologie-Abhandlungen 214(1–2): 353–358

Eder W (2008) Geoparks—promotion of earth sciences through geoheritage conservation, education and tourism. J Geol Soc India 72(2):149–154
El Difrawy MA, Runge MG, Moufti MR, Cronin SJ, Bebbington M (2013) A first hazard analysis of the Quaternary Harrat Al-Madinah volcanic field, Saudi Arabia. J Volcanol Geoth Res 267:39–46
El Hadi H, Tahiri A, Fernando Simancas J, Gonzalez-Lodeiro F, Azor A, Martinez-Poyatos D (2011) Geoheritage in Morocco: the Neoproterozoic Ophiolite of Bou Azzer (Central Anti-Atlas). Geoheritage 3(2):89–96
El Wartiti M, Malaki A, Zahraoui M, El Ghannouchi A, Di Gregorio F (2008) Geosites inventory of the northwestern tabular middle Atlas of Morocco. Environ Geol 55(2):415–422
Erfurt-Cooper P (2011) Geotourism in volcanic and geothermal environments: playing with fire? Geoheritage 3(3):187–193
Erfurt-Cooper P (2014) Volcanic tourist destinations. In: Eder W, Bobrowsky PT, Martínez-Frías J (eds) Geoheritage, geoparks and geotourism. Springer, Heidelberg, p 360
Erhartic B (2010) Geomorphosite assessment. Acta Geographica Slovenica-Geografski Zbornik 50(2):296–309
Errami E, Brocx M, Semeniuk V (2015) From geoheritage to geoparks—case studies from Africa and beyond. In: Eder W, Bobrowsky PT, Martínez-Frías J (eds) Geoheritage, geoparks and geotourism. Springer, Heidelberg, p 269
Fang SM, Tang JY (2010) Information technology and geological heritage management. In: Duserick F (ed) Ninth Wuhan international conference on e-business, Vols I-Iii. pp 1964–1970
Farsani NT, Coelho C, Costa C (2012) Geotourism and geoparks as gateways to socio-cultural sustainability in Qeshm Rural Areas, Iran. Asia Pac J Tourism Res 17(1):30–48
Farsani NT, Coelho COA, Costa CMM (2014a) Analysis of network activities in geoparks as geotourism destinations. Int J Tourism Res 16(1):1–10
Farsani NT, Coelho COA, Costa CMM, Amrikazemi A (2014b) Geo-knowledge management and geoconservation via geoparks and geotourism. Geoheritage 6(3):185–192
Fassoulas C, Mouriki D, Dimitriou-Nikolakis P, Iliopoulos G (2012) Quantitative assessment of geotopes as an effective tool for geoheritage management. Geoheritage 4(3):177–193
Feuillet T, Sourp E (2011) Geomorphological heritage of the Pyrenees National Park (France): assessment, clustering, and promotion of geomorphosites. Geoheritage 3(3):151–162
Fuertes-Gutierrez I, Fernandez-Martinez E (2012) Mapping geosites for geoheritage management: a methodological proposal for the regional park of Picos de Europa (Leon, Spain). Environ Manag 50(5):789–806
Gavrila IG, Anghel T (2013) Geomorphosites inventory in the Macin Mountains (South-Eastern Romania). GeoJournal Tourism Geosites 11(1):42–53
Ghazi JM, Olafsdottir R, Tongkul F, Ghazi JM (2013) Geological features for geotourism in the western part of Sahand Volcano, NW Iran. Geoheritage 5(1):23–34
Giusti C, Calvet M (2010) The inventory of French geomorphosites and the problem of nested scales and landscape complexity. Geomorphologie-Relief Processus Environ 2:223–244
Godoy LH, Sardinha DdS, Bertini RJ, Conceicao FTd, Roveri Cd, Moreira CA (2013) Geopark potential of Uberaba (MG): geodiversity and geoconservation. Sociedade & Natureza 25(2):395–410
Gogin IY, Vdovets MS (2014) Geosites of international significance in the UNESCO WHS Lena Pillars Nature Park (Sakha Republic, Russia). Geoheritage 6(3):173–184
Gordon JE (2012) Rediscovering a sense of wonder: geoheritage, geotourism and cultural landscape experiences. Geoheritage 4(1): 65–77
Gray M (2008) Geodiversity: developing the paradigm. Proc Geolo Assoc 119:287–298
Guijón R, Henriquez F, Naranjo JA (2011) Geological, geographical and legal considerations for the conservation of unique iron oxide and sulphur flows at El Laco and lastarria volcanic complexes, central andes, Northern Chile. Geoheritage 3(4):299–315
Guilbaud M-N, Siebe C, Agustin-Flores J (2009) Eruptive style of the young high-Mg basaltic-andesite Pelagatos scoria cone, southeast of México City. Bull Volcanol 71(8):859–880
Guilbaud M-N, Siebe C, Layer P, Salinas S (2012) Reconstruction of the volcanic history of the Tacambaro-Puruaran area (Michoacan, Mexico) reveals high frequency of Holocene monogenetic eruptions. Bull Volcanol 74(5):1187–1211
Harangi S (2014) Volcanic heritage of the Carpathian-Pannonian Region in Eastern-Central Europe. In: Erfurt-Cooper P (ed) Volcanic tourist destinations. Springer, Heidelberg, pp 103–124
Harmon B, Viles H (2013) Beyond geomorphosites: trade-offs, optimization, and networking in heritage landscapes. Environ Syst Decisions 33(2, Sp. Iss. SI):272–285
Hassan SS, Kowalczyk A, Einafshar A (2012) A tourism demand based method of geosites assessment on geotourism prioritization modeling: the case of Razavi Khorasan Province. J Hosp Manag Tourism 3(5):82–94
Henriques MH, Neto K (2015) Geoheritage at the Equator: selected geosites of Sao Tome Island (Cameron Line, Central Africa). Sustainability 7(1):648–667
Henriques MH, Tomaz C, Sa AA (2012) The Arouca Geopark (Portugal) as an educational resource: a case study. Episodes 35 (4):481–488
Hurtado H, Dowling R, Sanders D (2014) An exploratory study to develop a geotourism typology model. Int J Tourism Res 16 (6):608–613
Hwan KC (2011) The selection of geosites and making storytelling at Gangwon-Do DMZ Geopark. J Assoc Korean Photo-Geographers 21(1):117–134
Ieleniz M (2009) Geotope, geosite, geomorphosite. Ann Valahia University of Târgovişte, Geogr Ser 9:7–22
Ilies DC, Josan N (2009) Geosites-geomorphosites and relief. GeoJournal Tourism Geosites 3(1):78–85
Joyce B (2009) Geomorphosites and volcanism. In: Reynard E, Coratz P, Regolini-Bissing G (eds) Geomorphosites. Verlag Dr. Friedrich Pfeil, Germany, pp 175–188
Joyce E (2010) Australia's geoheritage: history of study, a new inventory of geosites and applications to geotourism and geoparks. Geoheritage 2(1–2):39–56
Kazanci N (2012) Geological background and three vulnerable geosites of the Kizilcahamam-Camlidere Geopark project in Ankara, Turkey. Geoheritage 4(4):249–261
Keever PJM, Zouros N (2005) Geoparks: celebrating earth heritage, sustaining local communities. Episodes 28(4):274–278
Kereszturi G, Németh K (2012) Monogenetic basaltic volcanoes: genetic classification, growth, geomorphology and degradation. In: Németh K (ed) Updates in volcanology—new advances in understanding volcanic systems. inTech Open, Rijeka, Croatia, pp 3–88. http://dx.doi.org/10.5772/51387
Kereszturi G, Németh K, Csillag G, Balogh K, Kovács J (2011) The role of external environmental factors in changing eruption styles of monogenetic volcanoes in a Mio/Pleistocene continental volcanic field in western Hungary. J Volcanol Geoth Res 201(1–4):227–240
Kereszturi G, Jordan G, Németh K, Doniz-Paez JF (2012) Syn-eruptive morphometric variability of monogenetic scoria cones. Bull Volcanol 74(9):2171–2185
Kereszturi G, Nemeth K, Cronin SJ, Procter J, Agustin-Flores J (2014) Influences on the variability of eruption sequences and style

transitions in the Auckland volcanic field, New Zealand. J Volcanol Geoth Res 286:101–115

Kervyn M, Ernst GGJ, Carracedo JC, Jacobs P (2012) Geomorphometric variability of "monogenetic" volcanic cones: evidence from Mauna Kea, Lanzarote and experimental cones. Geomorphology 136(1):59–75

Kwon D (2013) Sanin Kaigan Geopark in Japan based on geotourism. J Korean Geomorphol Assoc 20(2):103–116

Lansigu C, Bosse-Lansigu V, Le Hebel F (2014) Tools and methods used to represent geological processes and geosites: graphic and animated media as a means to popularize the scientific content and value of geoheritage. Geoheritage 6(2):159–168

Lim K (2014) A study of geotourism growth through recognition of geoeducation and geoconservation for the geoheritage. J Tourism Leisure Res 26(6):43–59

Liu SW, Tian MZ, Shi WQ (2010) Digital earth, a new approach for the geoconservation-case study of Hexigten Global Geopark, Inner Mongolia. In: Guo H, Wang C (eds) Sixth international symposium on digital earth: models, algorithms, and virtual reality

Lorenz V (1986) On the growth of maars and diatremes and its relevance to the formation of tuff rings. Bull Volcanol 48:265–274

Lutz H, Lorenz V (2013) Early volcanological research in the Vulkaneifel, Germany, the classic region of maar-diatreme volcanoes: the years 1774–1865. Bull Volcanol 75(8):1–16

Mansur KL, Carvalho IdS (2011) Characterization and valuation of the geological heritage identified in the Pero dune field, state of Rio de Janeiro, Brazil. Geoheritage 3(2):97–115

Manville V, Németh K, Kano K (2009) Source to sink: a review of three decades of progress in the understanding of volcaniclastic processes, deposits, and hazards. Sediment Geol 220:136–161

Martin U, Németh K (2004) Mio/Pliocene phreatomagmatic volcanism in the western Pannonian Basin. Geological Institute of Hungary, Budapest, pp 1–193

Martinez-Torres LM, Alonso J, Valle JM (2011) The upper Aptian-Lower Albian Amber deposit of the Penacerrada II Geosite (Basque-Cantabrian Basin, northern Spain): geological context and protection. Geoheritage 3(1):55–61

Mazumdar MK (2007) The 'geopark' initiative. Curr Sci 92(1):12

Miccadei E, Sammarone L, Piacentini T, D'Amico D, Mancinelli V (2014) Geotourism in the Abruzzo, Lazio and Molise National Park (central Italy): the example of Mount Greco and Chiarano Valley. GeoJournal Tourism Geosites 13(1):38–51

Moreira J (2012) Interpretative panels about the geological heritage—a case study at the Iguassu Falls National Park (Brazil). Geoheritage 4 (1–2):127–137

Moufti MR, Németh K (2013) The intra-continental Al Madinah volcanic field, western Saudi Arabia: a proposal to establish Harrat Al Madinah as the first volcanic geopark in the Kingdom of Saudi Arabia. Geoheritage 5(3):185–206

Moufti MR, Moghazi AM, Ali KA (2013a) Ar-40/Ar-39 geochronology of the Neogene-Quaternary Harrat Al-Madinah intercontinental volcanic field, Saudi Arabia: implications for duration and migration of volcanic activity. J Asian Earth Sci 62:253–268

Moufti MR, Németh K, El-Masry N, Qaddah A (2013b) Geoheritage values of one of the largest maar craters in the Arabian Peninsula: the Al Wahbah Crater and other volcanoes (Harrat Kishb, Saudi Arabia). Cent Eur J Geosci 5(2):254–271

Moufti MR, Németh K, El-Masry N, Qaddah A (2014) volcanic geotopes and their geosites preserved in an arid climate related to landscape and climate changes since the Neogene in northern Saudi Arabia: Harrat Hutaymah (Hai'il Region). Geoheritage. doi:10.1007/s12371-014-0110-3

Németh K (2010) Monogenetic volcanic fields: origin, sedimentary record, and relationship with polygenetic volcanism. Geol Soc Am Spec Pap 470:43–66

Németh K, Cronin SJ (2009) Volcanic structures and oral traditions of volcanism of western Samoa (SW Pacific) and their implications for hazard education. J Volcanol Geoth Res 186(3–4):223–237

Németh K, Kereszturi G (2015) Monogenetic volcanism: personal views and discussion. Int J Earth Sci 104(8):2131–2146

Németh K, Martin U (1999) Late Miocene paleo-geomorphology of the Bakony-Balaton Highland Volcanic Field (Hungary) using physical volcanology data. Zeitschrift für Geomorphologie 43(4):417–438

Németh K, Haller MJ, Siebe C (2011) Maars and scoria cones: the enigma of monogenetic volcanic fields. J Volcanol Geoth Res 201 (1–4):V–Viii

Németh K, Cronin SJ, Smith IEM, Flores JA (2012) Amplified hazard of small-volume monogenetic eruptions due to environmental controls, Orakei Basin, Auckland volcanic field, New Zealand. Bull Volcanol 74(9):2121–2137

Olafsdottir R, Dowling R (2014) Geotourism and geoparks-a tool for geoconservation and rural development in vulnerable environments: a case study from Iceland. Geoheritage 6(1):71–87

Orsi G, Vito MA, Isaia R (2004) Volcanic hazard assessment at the restless Campi Flegrei caldera. Bull Volcanol 66(6):514–530

Paez JD (2010) The cinder cones of Tenerife (Canary Islands, Spain): implications for volcanic hazard and volcanic risk. Scripta Nova-Revista Electronica de Geografia y Ciencias Sociales 14(324)

Panizza M (2001) Geomorphosites: concepts, methods and examples of geomorphological survey. Chin Sci Bull 46:4–6

Pelfini M, Bollati I (2014) Landforms and geomorphosites ongoing changes: concepts and implications for geoheritage promotion. Quaestiones Geographicae 33(1):131–143

Pellitero R, Jose Gonzalez-Amuchastegui M, Ruiz-Flano P, Serrano E (2011) Geodiversity and geomorphosite assessment applied to a natural protected area: the Ebro and Rudron Gorges Natural Park (Spain). Geoheritage 3(3):163–174

Pereira P, Pereira D (2010) Methodological guidelines for geomorphosite assessment. Geomorphologie-Relief Processus Environ 2:215–222

Petrovic MD, Vasiljevic DA, Vujicic MD, Hose TA, Markovic SB, Lukic T (2013) Global geopark and candidate—comparative analysis of Papuk Mountain Geopark (Croatia) and Fruska Gora Montain (Serbia) by using GAM Model. Carpathian J Earth Environ Sci 8(1):105–116

Pica A, Fredi P, Monte Md (2014) The Ernici Mountains geoheritage (central Apennines, Italy): assessment of the geosites for geotourism development. GeoJournal Tourism Geosites 14(2):193–206

Pulido Fernandez M, Lagar Timon D, Garcia Marin R (2014) Geosites inventory in the geopark Villuercas-Ibores-Jara (Extremadura, Spain): a proposal for a new classification. Geoheritage 6(1):17–27

Reynard E, Hoblea F, Cayla N, Gauchon C (2011) Iconic sites for Alpine geology and geomorphology rediscovering Heritage? Revue De Geographie Alpine-Journal of Alpine Research 99(1–4):188–202

Risso C, Németh K, Combina AM, Nullo F, Drosina M (2008) The role of phreatomagmatism in a Plio-Pleistocene high-density scoria cone field: Llancanelo volcanic field (Mendoza), Argentina. J Volcanol Geoth Res 169(1–2):61–86

Rocha J, Brilha J, Henriques MH (2014) Assessment of the geological heritage of Cape Mondego natural monument (central Portugal). Proc Geol Assoc 125(1):107–113

Ruban DA (2010) Quantification of geodiversity and its loss. Proc Geol Assoc 121(3):326–333

Runge MG, Bebbington MS, Cronin SJ, Lindsay JM, Kenedi CL, Moufti MRH (2014) Vents to events: determining an eruption event record from volcanic vent structures for the Harrat Rahat, Saudi Arabia. Bull Volcanol 76(3):1–16

Sanchez Cortez JL (2013) The geoparks as binding conservation sources: geodiversity, biodiversity and cultural heritage. Nat Conserv 6(1):46–53

Sang AK (2014) Research strategy on columnar joint in South Korea. J Korean Earth Sci Soc 35(7):501–517
Semeniuk V, Milne LA, Waterhouse H (2000) Contemporary and sub-recent pollen record for the Leschenault Inlet estuary: towards a palynological baseline. J Roy Soc W Aust 83(4):335–348
Sheth HC, Ray JS, Bhutani R, Kumar A, Awasthi N (2010) The latest (2008–09) eruption of Barren Island volcano, and some thoughts on its hazards, logistics and geotourism aspects. Curr Sci 98(5):620–626
Sohnyoungkwan, Sik WK, 권창우, 김련, Jeon YM (2009) Geoheritages and geomonitoring with special reference to Jeju Island. J Geol Soc Korea 45(6):751–770
Sik WK, Hoon YS, Sohnyoungkwan, 김련, Lee KC, Lim J-D (2013) Natural heritage values and diversity of geoheritages on Udo Island, Jeju Province. MUN HWA JAE—Ann Rev Cult Heritage Stud 46 (1):290–317
Stoppa F, Schiazza M (2013) An overview of monogenetic carbonatitic magmatism from Uganda, Italy, China and Spain: volcanologic and geochemical features. J S Am Earth Sci 41:140–159
Strba Lu, Rybar P, Balaz B, Molokac M, Hvizdak L, Krsak B, Lukac M, Muchova L, Tometzova D, Ferencikova J (2015) Geosite assessments: comparison of methods and results. Curr Issues Tourism 18(5):496–510
Tefogoum GZ, Dongmo AK, Nkouathio DG, Wandji P, Dedzo MG (2014) Geomorphological features of the Manengouba Volcano (Cameroon Line): assets for potential geopark development. Geoheritage 6(3):225–239
Thordarson T, Self S (1993) The Laki (Skaftar Fires) and Grimsvotn eruptions in 1783–1785. Bull Volcanol 55(4):233–263
Tomic N, Bozic S (2014) A modified geosite assessment model (M-GAM) and its application on the Lazar Canyon area (Serbia). Int J Environ Res 8(4):1041–1052
Ulrich G, Holm R, Wolfe E (1989) San Francisco volcanic field. In: Chapin C, Zidek J (eds) Field excursions to volcanic terranes in the western United States, Volume I: Southern Rocky Mountain region. pp 10–21
Valentine GA, Gregg TKP (2008) Continental basaltic volcanoes—processes and problems. J Volcanol Geoth Res 177(4):857–873
Vergniolle S (1996) Bubble size distribution in magma chambers and dynamics of basaltic eruptions. Earth Planet Sci Lett 140(1–4): 269–279
Vergniolle S, Brandeis G (1996) Strombolian explosions.1. A large bubble breaking at the surface of a lava column as a source of sound. J Geophys Research-Solid Earth 101(B9):20433–20447
Vergniolle S, Manga M (2000) Hawaiian and strombolian eruptions. In: Sigurdsson H, Houghton BF, McNutt SR, Rymer H, Stix J (eds) Encyclopedia of volcanoes. Academic Press, San Diego, pp 447–461
Vespermann D, Schmincke H-U (2000) Scoria cones and tuff rings. In: Sigurdsson H, Houghton BF, McNutt SR, Rymer H, Stix J (eds) Encyclopedia of volcanoes. Academic Press, San Diego, pp 683–694
Vujičić M, Vasiljević ĐA, Marković SB, Hose TA, Lukić T, Hadžić O, Janićević S (2011) Preliminary geosite assessment model (GAM) and its application on Fruška Gora Mountain, potential geotourism destination of Serbia. Acta Geographica Slovenica-Geografski Zbornik 51(2):361–377
Walker GPL (1993) Basaltic-volcano systems. Geol Soc Spec Publ 76:3–38
Wang L, Tian M, Wen X, Zhao L, Song J, Sun M, Wang H, Lan Y, Sun M (2014) Geoconservation and geotourism in Arxan-Chaihe Volcano Area, Inner Mongolia, China. Quatern Int 349:384–391
White JDL, Ross PS (2011) Maar-diatreme volcanoes: a review. J Volcanol Geoth Res 201(1–4):1–29
Woo KS, Sohn YK, Yoon S-H, Ahn US, Spate A (2013) Jeju Island Geopark—a volcanic wonder of Korea. In: Eder W, Bobrowsky PT, Martínez-Frías J (eds) Geoheritage, geoparks and geotourism. Springer, Heidelberg, p 88
Yan Z, Feng W, Cao W, Yang Q (2008) A Preliminary study on e-business system of World Geoparks in China. pp 798–802. 978-0-9800510-0-1
Yan Z, Ma X, Sha T (2009) The virtual display study on 3D Panorama in tourist Areas-Take Shilin World Geopark as an example. pp 229–232. 978-952-5726-02-2
Yan Z, Shan Z, Yan B, Li X (2010) Study of sustainable development on Hubei Geopark. pp 436–439. 978-1-935068-32-7
Yildirim T, Kocan N (2008) Evaluated for geotourism of Nevsehir Acigol Calderas Kalecitepe and Acigol Maars. Ege Universitesi Ziraat Fakultesi Dergisi 45(2):135–143
Zouros N (2004) The European geoparks network—geological heritage protection and local development. Episodes 27(3):165–171
Zouros N, Pavlides S, Soulakellis N, Chatzipetros A, Vasileiadou K, Valiakos I, Mpentana K (2011) Using active fault studies for raising public awareness and sensitisation on seismic hazard: a case study from Lesvos Petrified Forest Geopark, NE Aegean Sea, Greece. Geoheritage 3(4):317–327
박준형, Cheong D (2012) A quantitative evaluation model for geoheritage. J Geol Soc Korea 48(2):163–178
배수경, 추창오, Jang Y-D (2014) Volcanological features and activation plan as a geosite of Dodong and Jeodong seashore walkways in Ulleung Island. J Geol Soc Korea 50(1):71–89
진광민, Kim Y-S (2010) Study on development and value for geotourism of horizontal columnar joints at the Jeongja beach area, Ulsan and Eupchon beach area, Gyeongju. J Geol Soc Korea 46 (4):413–427
박종관 (2013) Morphological landscape of Patagonia and Atacama in the Andes based on geotourism. J Korean Geomorphol Assoc 20 (4):15–28
조규성, 박경진, 양우헌 (2014) Geotourism and educational utilization of the geosites in the Byeonsanbando National Park. J Geol Soc Korea 50(1):107–120

2 Geological Setting

In this chapter the intraplate volcanic fields of the Arabian Peninsula will be presented in a global context through comparison with other intraplate volcanic fields, focusing on those with special relevance to known and/or planned geoheritage, geoconservation and geotouristic programs and geoparks.

2.1 Cenozoic Volcanic Fields of the Arabian Peninsula

The Red Sea Rift forms an active deformation zone between the African and Arabian continental plates, stretching about 2000 km from NNW to SSE (Fig. 2.1). The rifting along the Red Sea started about 30 million years ago, leading to the separation of the Arabian Plate from Africa (Avni et al. 2012; Bohannon et al. 1989; Bosworth et al. 2004, 2005; Camp and Roobol 1992; Corti 2009; Ghebreab 1998; Girdler 1991; Zeyen et al. 1997). The evolution of the marginal areas in reference to the axis of the Red Sea Rift was complex and consisted of a combination of (1) extension along the Red Sea basin (Bellahsen et al. 2003; McGuire and Bohannon 1989; Voggenreiter et al. 1988; Wernicke 1985); (2) a pronounced continental collision between Arabia and Eurasia since the middle Miocene about 13 Ma (Dewey et al. 1986; McKenzie 1978); and (3) the development of left-lateral strike slip zones in the northwest margin of the Arabian micro-continent (Garfunkel 1981; Garfunkel and Ben-Avraham 2001; Garfunkel and BenAvraham 1996). This geodynamically complex situation provided mantle melting and shear that fed magma rising to the surface, especially in the western and northern margin of the Arabian plate, to form volcanic fields close to the plate boundaries (Bord and Bertrand 1995; Camp et al. 1987; Camp and Roobol 1989, 1991, 1992; Camp et al. 1991, 1992; Harlavan et al. 2002; Ibrahim and Al-Malabeh 2006; Kaliwoda et al. 2007; Moufti and Anonymous 2004; Moufti et al. 2012a, b). As a result of this intracontinental volcanism, a thick pile of sheet-like lava flows and associated networks of shield and fissure-fed volcanoes formed in the past 30 million years across various harrats.

These lava flow dominated fields are generally known locally as harrats (Fig. 2.2). The word "harrat" is the possessive form of the singular Arabic noun "harra", which means "stony area, volcanic country, lava field" (Wehr 1976); it is related to the adjective "harr", meaning "hot" (cross-referenced from Camp and Robol 1989). The term harrat is commonly used and understood as a synonym for volcanic field. The generation of names for specific harrats is commonly locality driven, and refers to a nearby settlement or geographical marker. In this respect, especially for harrats that cover large surface areas, the boundary between specifically named harrats is fairly undefined or at least *ad hoc*. Here we follow the traditional naming of specific harrats, and will express a harrat's potential volcanological significance through its association with a potential volcanic field, as defined in the volcanic literature (Barde-Cabusson et al. 2014; Brenna et al. 2012; Cimarelli et al. 2013; Connor and Conway 2000; Hernando et al. 2014; Le Corvec et al. 2013a, b; Németh 2010; Németh et al. 2011; Runge et al. 2014; Valentine and Gregg 2008).

Among these harrats, the largest (both in eruptive volume and surface area) is Harrat Ash Shaam (Fig. 1.1), which covers an area of 50,000 km^2 (Al Kwatli et al. 2012; Ibrahim et al. 2003; Ilani et al. 2001; Shaw et al. 2003; Trifonov et al. 2011; Weinstein et al. 2006). While Harrat Ash Shaam is the largest harrat in the Arabian post-30 Ma volcanic regions, the majority of its area is outside of the territory of the Kingdom of Saudi Arabia. In addition this harrat is difficult to access and no geoheritage studies have been performed on it; therefore, it is not included in this work. The largest harrats by occupied surface area, number of vents and

© Springer International Publishing Switzerland 2016
M.R. Moufti and K. Németh, *Geoheritage of Volcanic Harrats in Saudi Arabia*,
Geoheritage, Geoparks and Geotourism, DOI 10.1007/978-3-319-33015-0_2

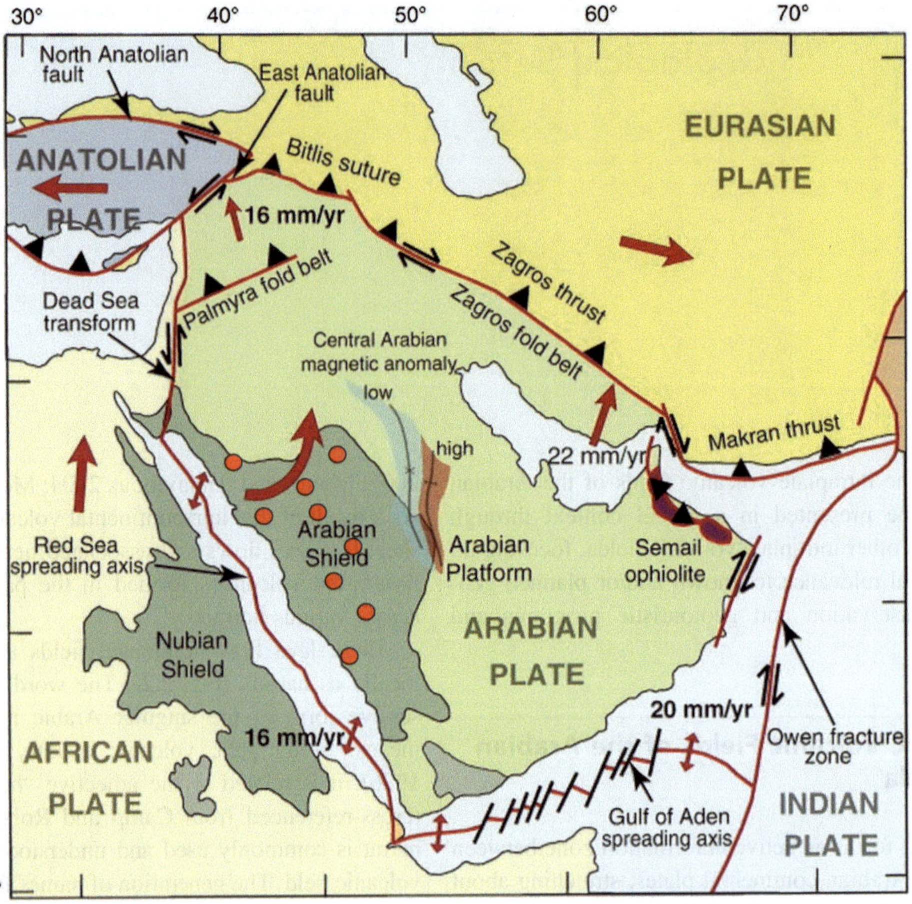

Fig. 2.1 General geotectonic framework of the Cenozoic harrats (*red dots*) of the Arabian Peninsula, concentrating volcanic fields mostly in the western region of the Kingdom of Saudia Arabia

Fig. 2.2 A typical view to a harrat in Saudi Arabia characteristic as a dark basalt-dominated rock desert such as shown in the picture from Harrat al Birk [17° 59′ 46.32″N; 41° 42′ 23.65″E]

estimated erupted volume in the territory of the Kingdom of Saudi Arabia are the Harrat Khaybar (20,564 km^2) and Harrat Rahat (19,830 km^2), which were both formed by a succession of volcanic eruptions at least 10 million years long that produced several hundreds of individual volcanoes ranging from those with basaltic compositions to rhyolites (Camp and Roobol 1989; Camp et al. 1991). The majority of the harrats produced volcanoes erupted from alkaline

Fig. 2.3 Peridotite lherzolite nodules (about 10 cm across) from the Harrat Kishb alkaline basaltic lava flows and pyroclastic deposits near Aslaj volcano [23° 14′ 18.16″N; 41° 16′ 1.26″E]

magmas, such as alkaline basalt, mugearite, benmoreite and trachyte typical for intracontinental volcanism (Camp and Roobol 1989, 1992; Camp et al. 1991; Coleman and Gregory 1983; Moufti et al. 2012a, b). In regions such as Harrat Kishb the volcanism was bimodal and produced phonolitic lava domes and colees (Camp et al. 1992; Coleman and Gregory 1983). Harrat Kishb is also the region where abundant deep seated mantle and deep crustal-derived xenoliths (Fig. 2.3) are known in the eruptive products, including various peridotite lherzolite nodules (McGuire 1988). A similar trend is documented at the Harrat Hutaymah in the NW edge of the region of harrats of the Arabian peninsula (Thornber 1990). The volcanic fields are erupted over old lithospheric fragments that are composed of various Precambrian rocks. The vents in most cases show strong parallel alignment with older, potentially reactivated, structural elements of the basement, often forming dorsal zones of volcanoes in the harrats, such as in the case of Harrat Rahat (Fig. 2.4). While patterns in the vent distribution are apparent (El Difrawy et al. 2013; Runge et al. 2014), the geological reasoning for them is not yet fully explained by detailed geological work and evidence.

2.2 Geological Setting of Harrat Rahat

The main subject of this book is to describe the geoheritage values of one of the largest harrats: Harrat Rahat. This volcanic region is located between the cities of Al Madinah, Jeddah and Makkah (Fig. 2.5) and, due to its good logistical position, its young volcanic landforms and numerous access points to its interior, it provides the perfect starting point to a develop volcanic geopark on the basis of defining its volcanic geoheritage sites.

The Rahat Volcanic Field defined by the Harrat Rahat consists of at least 500 individual volcanic edifices, many of them with typically complex edifice structures suggesting their prolonged volcanic activity (months to years) and common bimodal alkaline chemical nature (basaltic and trachytic) (Camp and Roobol 1989; Moufti et al. 2012a, b).

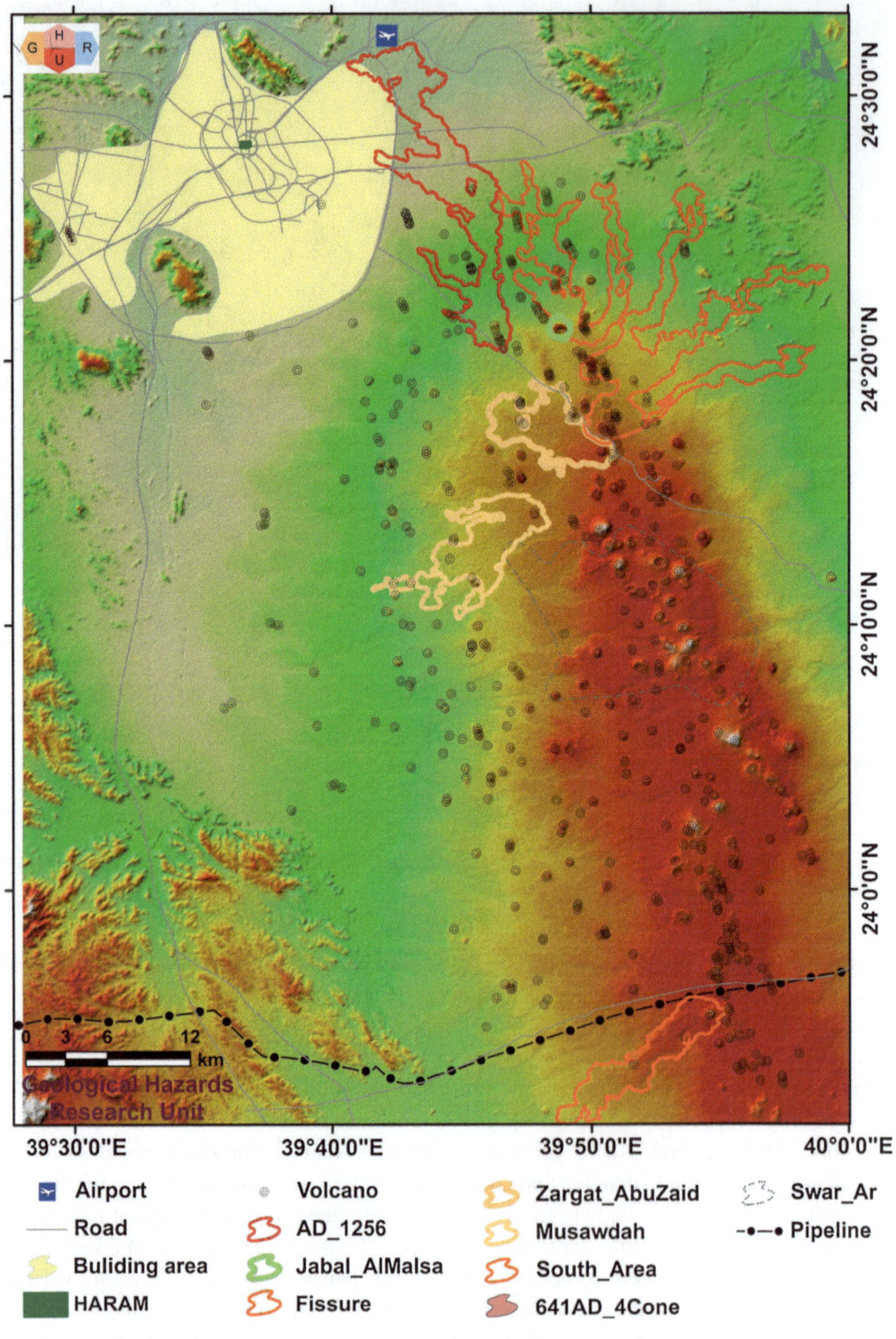

Fig. 2.4 Dorsal zone of the northern part of Harrat Rahat (Harrat al Madinah) shows high vent density and strong alignments of cones in small (edifice-scale) and large (field segment-scale) scales. Outlines of the youngest lava flows shown by thick coloured lines. The urban area with the harram in its centre is shown on the map in respect to their position to the volcanic field

The region located in the proximity of Al Madinah city is referred to as Harrat Al Madinah, and this region is extensively used as the main subject of this book. Due to logistical reasons, the dominant proportion of the detailed geological and stratigraphy information presented in this book comes from the Harrat Al Madinah region that then was extrapolated to the broader Harrat Rahat region.

The age of the volcanism of the Harrat Rahat was constrained dominantly by conventional K-Ar ages from whole rock samples from various lava flows (Brown et al. 1989; Camp and Roobol 1989; Coleman and Gregory 1983), as well as ages derived recently from Ar–Ar incremental heating techniques on groundmass separates from different volcanic lava flows from Harrat Al Madinah in the northernmost part of the Harrat Rahat (Moufti et al. 2012a, b).

Early K-Ar age determinations allowed the volcanic rocks to be divided into three major stratigraphic units formally defined as: Shawahit basalt ($\sim$10–2.5 Ma), Hammah

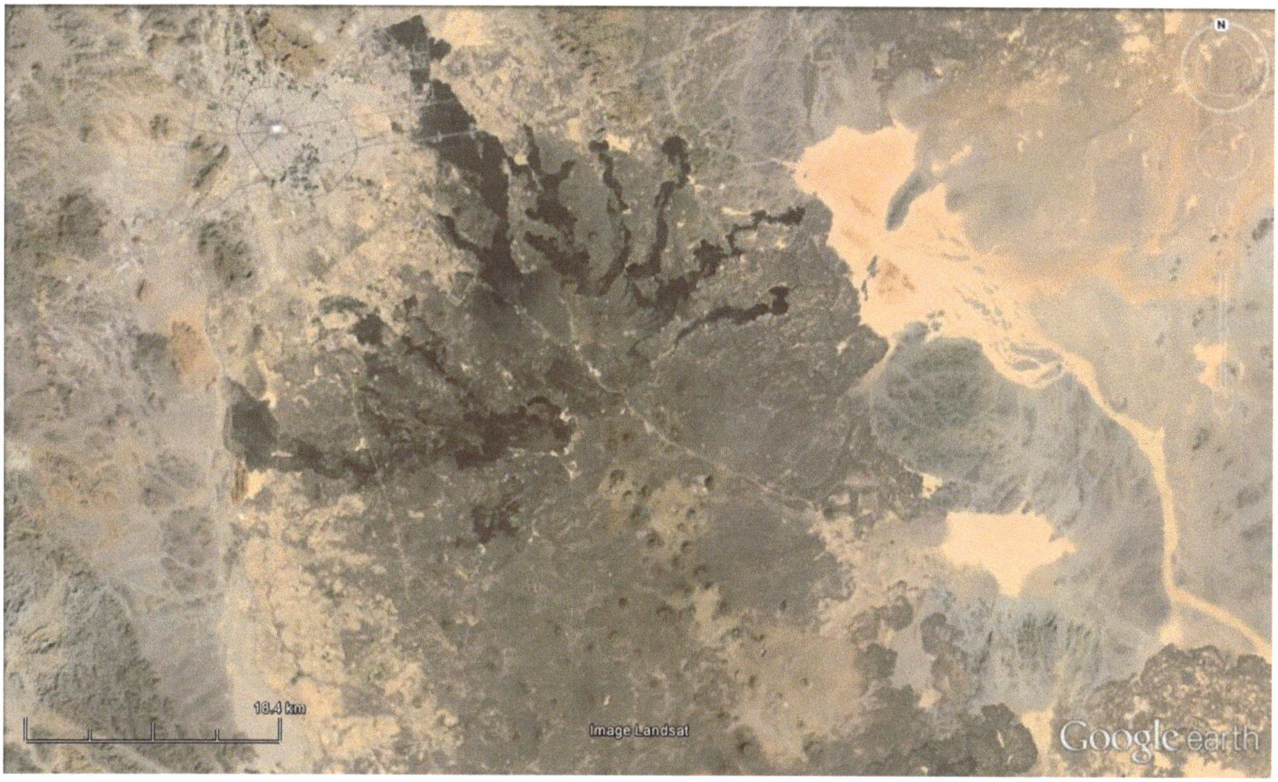

Fig. 2.5 GoogleEarth satellite image of the region of northern Harrat Rahat commonly referred as Harrat al Madinah. Al Madinah city is in the *top left corner* of the view

basalt (~2.5–1.7 Ma), and Madinah basalt (~1.7 Ma—Recent) (Moufti et al. 2012a, b). Geological field mapping, aided by K-Ar and recent Ar–Ar dating, subdivided the Madinah basalt into the lower and upper Madinah basalts (Moufti et al. 2012a, b). From a lithostratigraphical perspective, the lower Madinah basalt comprises three stratigraphic (mapping) sequences labelled Qm1 to Qm3 (Qm1: ~1.7–1.2 Ma; Qm2: ~1.2–0.9 Ma; Qm3: ~0.9–0.6 Ma); while the upper Madinah basalt includes four sequences defined as Qm4 to Qm7 (Qm4: ~0.6–0.3 Ma; Qm5: ~0.3 Ma—4500 B.P.; Qm6: ~4500–1500 B.P.; Qm7: ~1500 B.P.—1256 A.D.) (Camp and Roobol 1989).

Intensive geoheritage studies were conducted in the region of the Harrat Al Madinah where a proposal to establish a geopark, called the Harrat Al Madinah Volcanic Geopark, was put forward recently (Moufti and Németh 2013). This proposed geopark is located in the area covered by eruptive products grouped into these various lithostratigraphic units and represents the youngest volcanic episode of the Rahat Volcanic Field (RVF). There is an apparent northward migration of volcanic events, at least in the last 10 Ma, which has been linked by some workers to the age progression of the lithospheric up-doming of the Western Arabian Swell and the northernmost extremities of the larger, regional up-doming of the Afro-Arabian Dome (Almond 1986). This swell is inferred to be linked to the Ethiopean mantle plume, as its lobe reaches far north (Camp and Roobol 1992). Recent Ar–Ar dating has refined the previously proposed volcanic stratigraphy, providing evidence of far more evenly distributed volcanic events across the Harrat Rahat (Moufti et al. 2012a, b). Specifically, the longevity of volcanism in the northern section of the Harrat Rahat has been found to be greater than previously thought, suggesting less characteristic uni-directional migration of volcanic activity and challenging the idea of a fixed mantle plume over a steadily moving Arabian Plate as the source of the volcanism over the past 10 Ma (Moufti et al. 2012a, b). Instead, the NNW-trending distribution of the volcanic vents, i.e. nearly parallel to the Red Sea and its fault system, suggests that their origin is related to periodic extensional episodes along the reactivated Red Sea fault system (Moufti et al. 2012a, b).

The Harrat Rahat hosts numerous and diverse volcanic landforms that are well exposed and lack vegetation cover, offering a perfect site to see nearly unmodified volcanic landforms that are inferred to represent the syn-eruptive volcanic morphology of monogenetic volcanoes, as defined in (Kereszturi and Németh 2012). The most common volcanic landforms of the harrats in general are the basaltic scoria and lava spatter cones associated with pahoehoe and

Fig. 2.6 Lava domes are common in the Harrat Rahat and Khaybar. Many of them are silicic and they have a very hight aesthetic value such as the "White Mountains" of Harrat Khaybar [25° 39′ 28.36″N; 39° 58′ 13.69″E]

a'a lava fields. Many of them show a complex eruptive history with multiple craters and nested crater rims. Lava domes of mugearite, benmoreite and trachyte compositions are particularly common in the centre of the Harrat Rahat (Camp and Roobol 1989) and Harrat Khaybar (Camp et al. 1991) and form a spectacular scene of circular (in map view) and steep sided lava domes, many of them crowned with a characteristic solidified spine a few tens of metres above the main lava dome body, as seen at the "*White Mountains*" of Harrat Khaybar (Moufti and Nemeth 2014) (Fig. 2.6). In addition to constructional volcanic landforms, there are few, but large in diameter and crater depth, volcanic craters commonly surrounded by tuff rings or steep sided pyroclastic constructs closely resembling tuff cones. Such volcanic landforms are most common in the Harrat Hutaymah (Fig. 2.7a) in the north, but there are also fine examples at the Harrat Rahat (Fig. 2.7b), Harat Kishb (Fig. 2.7c) and Harrat Khaybar (Fig. 2.7d). Large areas are covered by trachytic tephra blankets that are the result of pyroclastic surges, block-and-ash flows, and fallout and commonly form extensive coloured surface regions in many of the harrats (Fig. 2.8).

The proposed Harrat Al Madinah Volcanic Geopark (HAMVG) can provide a holistic geoeducation and geoconservation program in a location where the diversity of intraplate dispersed volcanism in a long-lived volcanic field in an intra-continental region can be demonstrated. Volcanic fields can provide vital information on magma generation and ascent, on the style of volcanic eruptions and on the interaction between volcanism and the surrounding terrestrial basins in which the volcanoes erupted. This information can be related to the number and eruption styles of individual volcanoes (White 1991), the timing and frequency of eruptions (Conway et al. 1997, 1998; Kereszturi et al. 2011, 2013 Kiyosugi et al. 2010), the distribution pattern of volcanoes (Connor et al. 1992; Connor and Conway 2000; El Difrawy et al. 2013), and the relationship of the volcanoes to tectonic features, such as basins, faults, and rift zones (Connor 1987, 1990; Le Corvec et al. 2013a, b). Here we will present the major geoheritage values of the Harrat Al Madinah, followed by a comparative summary of geoheritage values of other harrats and demonstrate that the Saudi Arabian harrats can provide world-class sites to promote our understanding of one of the most common types of

Fig. 2.7 Negative volcanic landforms such as explosion craters and maars are most common in Harrat Hutaymah (**a** 26° 59′ 22.07″N; 42° 14′ 24.17″E]) but nice examples are known from the Harrat Kishb (**b** 22° 53′ 28.23″N; 41° 8′ 31.08″E), Harrat Rahat (**c** 24° 11′ 16.03″N; 39° 52′ 24.37″E) and Harrat Khaybar (**d** 25° 39′ 13.01″N; 39° 56′ 16.35″E) as well

Fig. 2.8 Light coloured ash plains of silicic tephra clearly visible on a GoogleEarth satellite image of the Harrat Khaybar. Top left corner of the map view is [25° 58′ 9.94″N; 39° 24′ 31.01″E]

volcanism on Earth and in the Solar System. In addition, the Saudi Arabain harrats together can form the geological basis to develop geoeducational programs for both the general public and the research community.

References

Al Kwatli MA, Gillot PY, Zeyen H, Hildenbrand A, Al Gharib I (2012) Volcano-tectonic evolution of the northern part of the Arabian plate in the light of new K-Ar ages and remote sensing: Harrat Ash Shaam volcanic province (Syria). Tectonophysics 580:192–207

Almond DC (1986) The relation of Mesozoic-Cenozoic volcanism to tectonics in the Afro-Arabian dome. J Volcanol Geoth Res 28:225–245

Avni Y, Segev A, Ginat H (2012) Oligocene regional denudation of the northern Afar dome: pre- and syn-breakup stages of the Afro-Arabian plate. Geol Soc Am Bull 124(11–12):1871–1897

Barde-Cabusson S, Gottsmann J, Marti J, Bolos X, Camacho AG, Geyer A, Planaguma L, Ronchin E, Sanchez A (2014) Structural control of monogenetic volcanism in the Garrotxa volcanic field (Northeastern Spain) from gravity and self-potential measurements. Bull Volcanol 76(1):1–13

Bellahsen N, Faccenna C, Funiciello F, Daniel JM, Jolivet L (2003) Why did Arabia separate from Africa? Insights from 3-D laboratory experiments. Earth Planet Sci Lett 216:365–381

Bohannon RG, Naeser CW, Schmidt DL, Zimmermann RA (1989) The timing of uplift, volcanism, and rifting peripheral to the Red Sea: a case for passive rifting? J Geophys Res 94(B2):1683–1701

Bord C, Bertrand H (1995) Geochemistry of recent alkaline volcanism from Saudi-Arabia—relevance to Red-Sea rifting. Comptes Rendus De L Academie Des Sciences Serie Ii 320(1):31–38

Bosworth W, Burke K, Anonymous (2004) Tectonic evolution of the Red Sea; integration and synthesis of outcrop, subsurface and radiometric data. International Geological Congress, Abstracts = Congres Geologique International, Resumes 32, Part 1:627–627

Bosworth W, Huchon P, McClay K (2005) The Red Sea and Gulf of Aden basins. J Afr Earth Sc 43(1–3):334–378

Brenna M, Cronin SJ, Smith IEM, Sohn YK, Maas R (2012) How small-volume basaltic magmatic systems develop: A case study from the Jeju Island Volcanic Field, Korea. J Petrol 53(5):985–1018

Brown GF, Schmidt DL, Huffan Jr AC (1989) Geology of the Arabian Peninsula, Shield area of western Saudi Arabia. US Geological Survey professional paper 560-A, pp 1–188

Camp VE, Roobol MJ (1989) The Arabian continental alkali basalt province; Part I, evolution of Harrat Rahat, Kingdom of Saudi Arabia; with Suppl. Data 89–04. Geol Soc Am Bull 101(1):71–95

Camp VE, Roobol MJ (1991) Topographic and volcanic asymmetry around the Red-Sea—constraints on rift models—comment. Tectonics 10(3):649–652

Camp VE, Roobol MJ (1992) Upwelling asthenosphere beneath Western Arabia and its regional implications. J Geophys Res Solid Earth 97(B11):15255–15271

Camp V, Hooper P, Roobol MJ, White DL (1987) The Madinah eruption, Saudi Arabia: Magma mixing and simultaneous extrusion of three basaltic chemical types. Bull Volcanol 49(2):489–508

Camp VE, Roobol MJ, Hooper PR (1991) The Arabian Continental Alkali Basalt Province. 2. Evolution of Harrats Khaybar, Ithnayn, and Kura, Kingdom of Saudi-Arabia. Geol Soc Am Bull 103(3):363–391

Camp VE, Roobol MJ, Hooper PR (1992) The Arabian Continental Alkali Basalt Province. 3. Evolution of Harrat Kishb, Kingdom of Saudi-Arabia. Geol Soc Am Bull 104(4):379–396

Cimarelli C, Di Traglia F, de Rita D, Gimeno Torrente D, Fernandez Turiel JL (2013) Space-time evolution of monogenetic volcanism in the mafic Garrotxa Volcanic Field (NE Iberian Peninsula). Bull Volcanol 75(11):1–18

Coleman RC, Gregory R (1983) Cenozoic volcanic rocks. US Geological Survey Professional Paper, pp 287–287

Connor CB (1987) Structure of the Michoacan-Guanajuato Volcanic Field, Mexico. J Volcanol Geoth Res 33(1–3):191–200

Connor CB (1990) Cinder cone clustering in the TransMexican Volcanic Belt: Implications for structural and petrologic models. J Geophys Res 95(B12):19,395–319,405

Connor CB, Conway FM (2000) Basaltic volcanic fields. In: Sigurdsson H (ed) Encyclopedia of volcanoes. Academic Press, San Diego, pp 331–343

Connor CB, Condit CD, Crumpler LS, Aubele JC (1992) Evidence of regional structural controls on vent distribution—Springerville Volcanic Field, Arizona. J Geophys Res Solid Earth 97 (B9):12349–12359

Conway FM, Ferrill DA, Hall CM, Morris AP, Stamatakos JA, Connor CB, Halliday AN, Condit C (1997) Timing of basaltic volcanism along the Mesa Butte Fault in the San Francisco Volcanic Field, Arizona, from Ar-40/Ar-39 dates: Implications for longevity of cinder cone alignments. J Geophys Res Solid Earth 102 (B1):815–824

Conway FM, Connor CB, Hill BE, Condit CD, Mullaney K, Hall CM (1998) Recurrence rates of basaltic volcanism in SP cluster, San Francisco volcanic field, Arizona. Geology 26(7):655–658

Corti G (2009) Continental rift evolution: from rift initiation to incipient break-up in the Main Ethiopian Rift, East Africa. Earth Sci Rev 96(1–2):1–53

Dewey JF, Hempton MR, Kidd WSF, Saroglu F, Sengör AMC (1986) Shortening of continental lithosphere: the neotectonics of Eastern Anatolia-a young collision zone. In: Coward MP, Ries AC (eds) Collision zone tectonics. Geological Society, London, London, pp 3–36

El Difrawy MA, Runge MG, Moufti MR, Cronin SJ, Bebbington M (2013) A first hazard analysis of the Quaternary Harrat Al-Madinah volcanic field, Saudi Arabia. J Volcanol Geoth Res 267:39–46

Garfunkel Z (1981) Internal structure of the Dead Sea leaky transform (rift) in relation to plate kinematics. Tectonophysics 80:81–108

Garfunkel Z, BenAvraham Z (1996) The structure of the Dead Sea basin. Tectonophysics 266(1–4):155–176

Garfunkel Z, Ben-Avraham Z (2001) Basins along the Dead Sea Transform. In: Ziegler PA, Cavazza W, Robertson AHF, CrasquinSoleau S (eds) Peri-Tethys Memoir 6: Peri-Tethyan Rift/Wrench basins and passive margins, pp 607–627

Ghebreab W (1998) Tectonics of the Red Sea region reassessed. Earth Sci Rev 45(1–2):1–44

Girdler RW (1991) The Afro-Arabian rift system—an overview. Tectonophysics 197(2–4):139–153

Harlavan Y, Weinberger R, Ilani S, Tarawneh K, Rabba I, Steinitz G, Anonymous (2002) K-Ar ages, Pb isotopes and geochemistry of basalts from Ash Shaam volcanic field, NE Jordan. Geochimica Et Cosmochimica Acta 66(15A):311–311

Hernando IR, Franzese JR, Llambias EJ, Petrinovic IA (2014) Vent distribution in the Quaternary Payun Matru Volcanic Field, western Argentina: Its relation to tectonics and crustal structures. Tectonophysics 622:122–134

Ibrahim KM, Al-Malabeh A (2006) Geochemistry and volcanic features of Harrat El Fahda: a young volcanic field in northwest Arabia, Jordan. J Asian Earth Sci 27(2):147–154

Ibrahim KM, Tarawneh K, Rabba I (2003) Phases of activity and geochemistry of basaltic dike systems in northeast Jordan parallel to the Red Sea. J Asian Earth Sci 21(5):467–472

Ilani S, Harlavan Y, Tarawneh K, Rabba I, Weinberger R, Ibrahim K, Peltz S, Steinitz G (2001) New K-Ar ages of basalts from the Harrat Ash Shaam volcanic field in Jordan: implications for the span and duration of the upper-mantle upwelling beneath the western Arabian plate. Geology 29(2):171–174

Kaliwoda M, Altherr R, Meyer H-P (2007) Composition and thermal evolution of the lithospheric mantle beneath the Harrat Uwayrid, eastern flank of the Red Sea Rift (Saudi Arabia). Lithos 99(1–2): 105–120

Kereszturi G, Németh K (2012) Monogenetic basaltic volcanoes: genetic classification, growth, geomorphology and degradation. In: Németh K (ed) Updates in volcanology—new advances in understanding volcanic systems. inTech Open, Rijeka, Croatia, pp 3–88 [http://dx.doi.org/10.5772/51387]

Kereszturi G, Németh K, Csillag G, Balogh K, Kovács J (2011) The role of external environmental factors in changing eruption styles of monogenetic volcanoes in a Mio/Pleistocene continental volcanic field in western Hungary. J Volcanol Geoth Res 201(1–4):227–240

Kereszturi G, Nemeth K, Cronin SJ, Agustin-Flores J, Smith IEM, Lindsay J (2013) A model for calculating eruptive volumes for monogenetic volcanoes—Implication for the Quaternary Auckland Volcanic Field, New Zealand. J Volcanol Geoth Res 266:16–33

Kiyosugi K, Connor CB, Zhao D, Connor LJ, Tanaka K (2010) Relationships between volcano distribution, crustal structure, and P-wave tomography: an example from the Abu Monogenetic Volcano Group, SW Japan. Bull Volcanol 72(3):331–340

Le Corvec N, Menand T, Lindsay J (2013a) Interaction of ascending magma with pre-existing crustal fractures in monogenetic basaltic volcanism: an experimental approach. J Geophys Res Solid Earth 118(3):968–984

Le Corvec N, Spoerli KB, Rowland J, Lindsay J (2013b) Spatial distribution and alignments of volcanic centers: clues to the formation of monogenetic volcanic fields. Earth Sci Rev 124:96–114

McGuire AV (1988) Petrology of mantle xenoliths from Harrat Al Khisb—the mantle beneath Western Saudi Arabia. J Petrol 29 (1):73–92

McGuire AV, Bohannon RG (1989) Timing of mantle upwelling; evidence for a passive origin for the Red Sea Rift. J Geophys Res Solid Earth 94:1677–1682

McKenzie D (1978) Active tectonics of the Alpine-Himalayan belt, the Aegean Sea and surrounding regions. Geophys J Roy Astron Soc 55:217–254

Moufti MR, Anonymous (2004) Sr-Nd isotopic and geochemical evidence for a mantle plume origin of Harrat Khaybar intraplate volcanic rocks, Kingdom of Saudi Arabia. International Geological Congress, Abstracts = Congres Geologique International, Resumes 32, Part 2:1024–1024

Moufti MR, Nemeth K (2014) The White Mountains of Harrat Khaybar, Kingdom of Saudi Arabia. Int J Earth Sci 103(6):1641–1643

Moufti MR, Németh K (2013) The intra-continental Harrat Al Madinah Volcanic Field, Western Saudi Arabia: a proposal to establish Harrat Al Madinah as the first volcanic geopark in the Kingdom of Saudi Arabia. Geoheritage 5(3):185–206

Moufti MR, Moghazi AM, Ali KA (2012a) 40Ar/39Ar geochronology of the Neogene-Quaternary Harrat Al-Madinah intercontinental Volcanic Field, Saudi Arabia: implications for duration and migration of volcanic activity. J Asian Earth Sci 62:253–268

Moufti MR, Moghazi AM, Ali KA (2012b) Geochemistry and Sr-Nd-Pb isotopic composition of the Harrat Al-Madinah Volcanic Field, Saudi Arabia. Gondwana Res 21(2–3):670–689

Németh K (2010) Monogenetic volcanic fields: origin, sedimentary record, and relationship with polygenetic volcanism. In: Canon-Tapia E, Szakacs A (eds) What is a volcano?. Geological Society of America, Boulder, Colorado, pp 43–66

Németh K, Haller MJ, Siebe C (2011) Maars and scoria cones: the enigma of monogenetic volcanic fields. J Volcanol Geoth Res 201 (1–4):V–Viii

Runge MG, Bebbington MS, Cronin SJ, Lindsay JM, Kenedi CL, Moufti MRH (2014) Vents to events: determining an eruption event record from volcanic vent structures for the Harrat Rahat, Saudi Arabia. Bull Volcanol 76(3):1–16

Shaw JE, Baker JA, Menzies MA, Thirlwall MF, Ibrahim KM (2003) Petrogenesis of the largest intraplate volcanic field on the Arabian plate (Jordan): a mixed lithosphere-asthenosphere source activated by lithospheric extension. J Petrol 44(9):1657–1679

Thornber CR (1990) Geologic map of Harrat Hutaymah, with petrologic classification and distribution of ultramafic inclusions, Saudi Arabia. In: US Geological Survey, Reston, VA, Reston, VA, United States (USA), pp 0–1 sheet

Trifonov VG, Dodonov AE, Sharkov EV, Golovin DI, Chernyshev IV, Lebedev VA, Ivanova TP, Bachmanov DM, Rukieh M, Ammar O, Minini H, Al Kafri AM, Ali O (2011) New data on the late Cenozoic basaltic volcanism in Syria, applied to its origin. J Volcanol Geoth Res 199(3–4):177–192

Valentine GA, Gregg TKP (2008) Continental basaltic volcanoes; processes and problems. J Volcanol Geoth Res 177(4):857–873

Voggenreiter W, Hötzl H, Jado A (1988) Red Sea related history of extension and magmatism in the Jizan area (southwest Saudi Arabia): indication for simple shear during Red Sea rifting. Geol Rundsch 77:257–274

Wehr H (1976) A dictionary of modern written Arabic [*An enlarged and improved version of 'Arabisches Wörterbuch für die Schriftsprache der Gegenwart' and includes the Contents of the 'Supplement zum Arabischen Wörterbuch für die Schriftsprache der Gegenwart by Otto Harrauowltz, Wiesbaden 1961, 1966, 1971*]. Spoken Language Services, Inc., New York, p 492. Third Edition [ISBN 0-87950-001-8]

Weinstein Y, Navon O, Altherr R, Stein M (2006) The role of lithospheric mantle heterogeneity in the generation of Plio-Pleistocene alkali basaltic suites from NW Harrat Ash Shaam (Israel). J Petrol 47(5):1017–1050

Wernicke B (1985) Uniform-sense normal simple-shear of the continental lithosphere. Can J Earth Sci 22:108–125

White JDL (1991) The depositional record of small, monogenetic volcanoes within terrestrial basins. Spec Pub Soc Econ Paleontol Mineral 45:155–171

Zeyen H, Volker F, Wehrle V, Fuchs K, Sobolev SV, Altherr R (1997) Styles of continental rifting: crust-mantle detachment and mantle plumes. Tectonophysics 278(1–4):329–352

3 Harrat Rahat: The Geoheritage Value of the Youngest Long-Lived Volcanic Field in the Kingdom of Saudi Arabia

3.1 Introduction

Harrat Rahat is a volcanic field that consists of over 500 individual volcanoes (Fig. 3.1), many of them with multiple vents forming compound edifices (Camp and Roobol 1989; Coleman and Gregory 1983; El Difrawy et al. 2013; Moufti et al. 2013a). Harrat Rahat was formed over the past 10 million of years (Moufti et al. 2013a), and it is still considered to be an active volcanic region as it has had at least two historic eruptions (Camp et al. 1987; Moufti et al. 2013a). The volcanic field consists of extensive lava fields (Murcia et al. 2014) and various types of volcanic cones and explosion craters (Camp et al. 1991; El Difrawy et al. 2013; Moufti and Hashad 2005; Moufti et al. 2011), each of them is perfectly exposed due to the arid climate and lack of vegetation, and many of them are relatively easy to access (Fig. 3.2). The field is located nearby one of the holiest cities of Islam—Al Madinah—and also hosts the youngest volcanoes in the Kingdom of Saudi Arabia, which have historical and cultural significance (Fig. 3.1). Harrat Al Madinah is the northern part of the Harrat Rahat and the best studied in the Harrat Rahat. The distinction between Harrat Rahat and Harrat Al Madinah is loosely constrained and it has a traditional and geographic connotation rather than geological reasoning. In a similar way, different parts of Harrat Rahat have local names that refer to nearby settlements or other geographical features.

In this chapter we will present a detailed summary of the geoheritage value of the geological features that form the backbone of the geoheritage of Harrat Rahat. The most extensive geoheritage research in Saudi Arabia has been undertaken in the Harrat Al Madinah in the northern part of Harrat Rahat, and that is the basis of a proposal to establish the Harrat Al Madinah Volcanic Geopark. As in other harrats in the Kingdom, the geoheritage research is rather fragmental so far; in subsequent chapters we will provide a brief summary of the geoheritage value associated with other harrats. In describing subsequent harrats we will refer heavily back to the identified geoheritage value of the Harrat Rahat, which will provide a firm scientific basis to justify the high geoheritage value of all the harrats of the Kingdom of Saudi Arabia. The harrats could thus be promoted as a continent-scale world heritage site on the basis of the universal value of observing and studying volcanism.

UNESCO promotes conservation of geological and geomorphological heritage through protection of world heritage sites and development of educational programs under the umbrella of geoparks (Dowling 2011; Farsani et al. 2011; Gordon 2012; Henriques et al. 2011; Joyce 2010). In this chapter we identify significant volcanic features that could be organized and promoted as the first geopark, the Al Madinah Volcanic Geopark in the Kingdom of Saudi Arabia (Moufti and Németh 2013a). The Harrat Al Madinah Volcanic Field has numerous volcanic geosites (Moufti and Németh 2013a, b, c) relevant to broadening our understanding of the evolution of volcanic fields dominated by Hawaiian and Strombolian style volcanic cones and lava fields (Kereszturi and Németh 2012a, b).

The proposed geopark includes the location of the last historically erupted volcanoes in the Arabian Peninsula (Moufti and Németh 2013a; Moufti et al. 2013b). An historic eruption site in the proximity of Al Madinah City formed a chain of lava spatter and scoria cones formed in a 52 day-long eruption in 1256 AD (Fig. 3.3). This eruption site is located about just 10 km SE of modern Al Madinah city (Fig. 3.1). The violent eruption formed a $\sim$2 km long NW-SE-aligned fissure that produced at least seven volcanic edifices with multiple vents that is now a globally unique volcanic landscape with easy access from a major city of Saudi Arabia. Any geoeducational and geoconservation program designed or proposed for this region in the future must take this location as the core of the project (Moufti and Németh 2013a, b, c).

© Springer International Publishing Switzerland 2016
M.R. Moufti and K. Németh, *Geoheritage of Volcanic Harrats in Saudi Arabia*,
Geoheritage, Geoparks and Geotourism, DOI 10.1007/978-3-319-33015-0_3

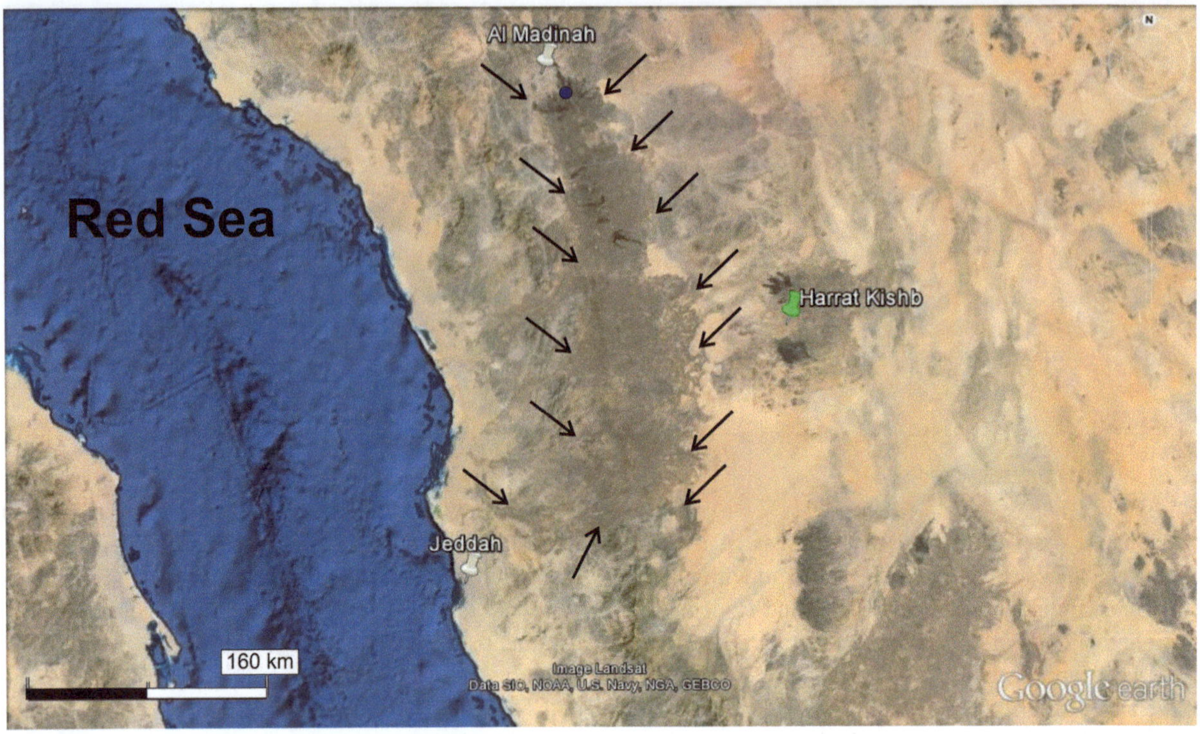

Fig. 3.1 Overview map of Harrat Rahat on GoogleEarth image. *Arrows* outline the boundary of Harrat Rahat. *Blue dot* marks the 1256 AD historic eruption site near Al Madinah City

Fig. 3.2 General view of Harrat Rahat with cones and flow fields [24° 20′ 14.91″N; 39° 51′ 6.33″E]

Fig. 3.3 Overview of the 1256 AD eruption site from the SE [24° 20′ 35.04″N; 39° 46′ 38.49″E]

Harrat Rahat consists of excellent geotopes that illustrate fine details of explosive and effusive volcanism of monogenetic volcanic fields. Thus this is one of the most accessible places on Earth to see the geological context of the birth, evolution and erosion of lava spatter and scoria cone complexes and their associated lava flow fields.

Because Harrat Al Madinah is located so near to Al Madinah city the proposed geopark is easily accessible through highways (and by train in the near future) and it would provide significant economic benefit to Al Madinah city. The park could provide a cost-effective volcanic geoeducation program to pilgrims who are in the city visiting the holy sites.

Through the creation of a world network of natural parks with significant geological features, labelled UNESCO Geoparks, UNESCO promotes conservation of our geological heritage (Dowling 2011; Erfurt-Cooper 2011; Farsani et al. 2011; Joyce 2010). The first step in developing a geopark is to identify geotopes, geosites and geomorphosites which are the key geological features in a region that are easy to access, significant in the global geological sense and that could potentially serve as a basis for broader geoconservation projects (Deraman et al. 2010; Moufti et al. 2013c; Petrovic et al. 2013). Volcanic geoparks are increasingly popular projects worldwide and play a substantial role in geohazard education, including facilitating the dissemination of current research results on the volcanic processes that the ever increasing human society faces (Erfurt-Cooper 2011).

In addition, volcanic geoparks can serve as a geotouristic base that can generate significant economic benefit. Geosites, geomorphosites and geotopes are the smallest "units" of intact geological features that are identifiable through their uniqueness or because they are graphic examples of specific volcanic phenomena or form a vital landscape representative of a specific volcanic processes (Armiero et al. 2011; Erikstad 2013; Fassoulas et al. 2012).

Here we identify significant volcanic features that bear not only regional, but global, volcanic value in a confined area that could be organized and promoted as the first volcanic geopark in the Kingdom of Saudi Arabia: the Al Madinah Volcanic Geopark (Moufti and Németh 2013a). Harrat Al Madinah has many volcanic geosites including the last historically erupted volcanoes in Arabia (Camp and Roobol 1989). Overall, the proposed geopark would provide significant economic benefit to the nearby city of Al Madinah. Pilgrims arrive from every corner of the globe,

Fig. 3.4 The 1256 AD eruption site on a LiDAR image shows the complex nested scoria cone structure of the eruption site. Highway is in the *bottom left corner* of the image about 600 m from the foothill of the cone complex. The top left corner of the map view is about [24° 21′ 39.72″N; 39° 45′ 57.75″E]

including countries where volcanic hazard is an everyday aspect of life (e.g. Indonesia); therefore, the proposed geopark would serve as a significant geoeducational hot spot (Moufti and Németh 2013a, b, c).

A major geotope with distinct geosites/geomorphosites has been selected to demonstrate the diversity of volcanic phenomena associated with the intraplate volcanism of the Harrat Al Madinah. Hawaiian to Strombolian type eruptions created lava spatter and scoria cones (Kereszturi and Németh 2012a, b) visible from major highways, allowing visitors to stop near the AD 1256 historic eruption site just 10 km SE of Al Madinah (Fig. 3.4). The 52 day-long eruption formed a ~2 km long NW-SE-aligned fissure which emitted mainly a'a lava flows and lava spatter-dominated pyroclastic cones (Camp et al. 1987) (Fig. 3.5).

At least seven vents have been identified, which made nested lava spatter cones (Camp et al. 1987; Murcia et al. 2013, 2014). The main central cones had more energetic explosive eruptive episodes that generated pyroclastic fall deposits, forming an ash-plain (Fig. 3.6) (Murcia et al. 2013). The vents are inferred to be hosted lava lakes and lava lake outbreaks initiated crater wall collapses, as traced on circular fissures along the crater margins (Murcia et al. 2014). The growth of the individual cones was repeatedly interrupted by lava flow outbreaks in the tip of the fissures by rafting away large pieces of the cones that were subsequently rehealed, resulting in a nested and complex volcano morphology (Murcia et al. 2014). The geotopes form the 1256 AD eruption sites as part of the Northern Harrat Rahat (or Harrat Al Madinah) is probably the best exposed and accessible site on Earth to show the diversity of volcanic features a fissure-eruption can produce (Moufti and Németh 2013a).

The recent increased seismic activity in 2009 in the region just north of Harrat Rahat in the Harrat Lunayyir region (Duncan and Al-Amri 2013; Hansen et al. 2013; Mukhopadhyay et al. 2013; Pallister et al. 2010; Zahran et al. 2009; Zobin et al. 2013), also justifies the establishment of an educational site that could play a significant role in the dissemination of scientific knowledge to the public, which could help the population better understand the potential outcome of any volcanic unrest the region may face (Moufti and Németh 2013a; Moufti et al. 2013b, c, d and e).

An historic review of seismic and volcanic events in the Arabian Peninsula, based on English translations of original documents, reveals that an earthquake occurred in 641 AD that destroyed houses in Al-Madinah (Ambraseys et al. 1994). It has been suggested that this earthquake is linked with a volcanic event outside of Harrat Rahat that occurred a year before, in 640 AD (Ambraseys et al. 1994). The location of this event is generally accepted to be a chain of four small cones west of Al-Madinah City (Camp and Roobol 1989), but on further examination the evidence justifying these four cones as the site of the 641 AD eruption is lacking (Moufti et al. 2013b). This volcanic event is associated with one or both of the following eruptions mentioned in historic

Fig. 3.5 The 1256 AD eruption site fissure aligned nature obvious feature visible for an untrained eyes. View looking toward NW

Fig. 3.6 Ash plain around the 1256 AD cones. View is looking toward the 1256 AD cones in the background from the point of about [24° 20′ 37.33″N; 39° 46′ 54.83″E]

records and occurred near to Tabuk (about 300 km NW of Al-Madinah City): the Hala'l-'Ishqa (27.58° N, 36.80° E) and/or Hala'l-Badr (27.25° N–37.20° E) (Ambraseys et al. 1994) in the Harrat Uwayrid (Fig. 3.1). Indeed there are young volcanic landforms located in this region judging from their morphological appearance but their historic age is questionable.

Camp and Roobol (1989) report references to a volcanic eruption in 641 AD located in the vicinity of Al Madinah that were reported in the manuscript of "Khulasaf Al-Wafa" which was written in 1568 AD by Nour Al-Dian B. Al-Samhoudy, commonly identified as the historian of Al-Madinah. Interestingly, Camp and Roobol (1989) agree with a report connecting this event to a volcanic eruption in 641 AD associated with a specific harrat called Harrat Layla (Simkin and Siebert 1994). Confusingly, Harrat Layla has been mentioned as the location of a fire (eruption?) in 640 AD—not in 641 AD—to where Umar, the ruler of Al-Madinah, ordered a man to go out, but in the meantime the "fire" was gone (Juynboll 1989a, b), suggesting a short-lived event at a distance from Al Madinah that could have been travelled in a single day (e.g. <100 km). As a conclusion, the eruption of 641 AD and its location are poorly constrained; however there is no doubt that the four small scoria cones that are located about 13 km to the SW from the Holy Mosque are very young cones (Fig. 3.7). These cones are the likely locations of a young volcanic event that could be the result of the 641 AD eruption (Moufti et al. 2013b).

3.2 Volcano Types and the Geoheritage Value of the Harrat Rahat

Harrat Rahat is one of the most diverse volcanic regions of the Arabian Peninsula in respect of the presence of well-preserved, young volcanic landforms and their eruptive products. While the Harrat Rahat is viewed as an intracontinental volcanic field with numerous monogenetic (short lived and small volume) volcanoes, its volcanological diversity is far greater from that. The most extensive volcanic features of the region are the various types of lava fields (Murcia et al. 2014). Many of the lava fields are associated clearly with point sources such as scoria and/or spatter cones or they have emerged along fissures defined by some sort of lineaments of relatively small size of cones. The majority of the lava flows are partially confined forming narrow branches of flows following gentle sloping low rimmed valleys (Fig. 3.8) (Murcia et al. 2014). It seems that the lava flow distribution has been strongly controlled by the landscape inundated by successive flow units gradually shifting younger flows side by side. As a result, a

Fig. 3.7 641 AD cones on GoogleEarth satellite image [24° 24′ 42.82″N; 39° 29′ 50.98″E]

Fig. 3.8 Confined lava flows occupy narrow valley systems in the Northern Harrat Rahat [24° 22′ 38.38″N; 39° 55′ 37.12″E]

characteristic flow pattern can be seen in many places, where older lava flows acted as obstacles for younger flows, especially in the northern side of the Harrat Rahat. Lava flows that outpoured along north to south aligned fissures tend to form distinct lava lobes from the north to south trending dorsal ridge of the Harrat Rahat.

The majority of the lava flows fields are transitional types, representing surface textures carry features typical of aa lavas that composed of pieces of broken partially developed cooling lava crusts (Fig. 3.9), commonly defined as rubbly or slabby pahoehoe (Murcia et al. 2014) similar to those flow fields documented in Cameroon (Suh et al. 2011; Wantim et al. 2011), Deccan in India (Bondre and Hart 2008; Duraiswami et al. 2003, 2014) or in Krafla in Iceland (Rossi 1997). The lava fields commonly engulf obstacles such as pre-flow cones (Fig. 3.10). In medial to distal areas, the lava flow fields are commonly littered by pieces of rafted cone material as a sign that the flows might have either emitted in a time when their source cone gradually collapsed or the flow itself bulldozed through pre-existing older cones (Németh et al. 2011; Riggs and Duffield 2008).

The most common types of volcanic edifices in the Harrat Rahat are the scoria cones and spatter cones. While no systematic study has been done on their morphometric parameters some preliminary study documented that their parameters range from the full spectrum of sizes known from

Fig. 3.9 Slabby pahoehoe in Harrat Rahat [24° 25′ 5.73″N; 39° 51′ 4.40″E]

Fig. 3.10 Engulfed pre-flow cone in the Northern Harrat Rahat [24° 22′ 36.68″N; 39° 49′ 26.03″E]

such cones on Earth (Kereszturi and Németh 2012a, b). There are a large number of relatively small lava spatter cones closely resembling large hornitos (Wentworth and Macdonald 1953) many of them with very steep slope angles (Fig. 3.11) (Moufti et al. 2013e).

On the other hand, very large scoria cones are known mostly from the dorsal zone of the Harrat Rahat (Fig. 3.12), that are closely resembling small stratovolcanoes with complex pyroclastic stratigraphy, suggesting their longer activity and larger eruptive volumes as a scoria cone generally considered for (Kereszturi and Németh 2012a, b). In this respect the larger scoria cones are better to view as long lived, small-to-medium sized stratovolcanoes, similar in eruption style and size as the active Cerro Negro in Nicaragua (Hill et al. 1998; McKnight and Williams 1997; Roggensack et al. 1997). In general the majority of the scoria cones of Harrat Rahat are dominated by pyroclastic deposits and demonstrate evidence of intense heat effect, agglutination, welding and accumulation of pyroclasts typical of eruption through vigorous lava fountain events. In many cases, the cones have well-preserved crater rims composed of welded pyroclasts (Fig. 3.13). These features together indicate that the eruptions must have been dominated by lava fountaining that produced spatter-like pyroclasts agglutinated and welded together upon landing (Head and Wilson 1989; Kereszturi and Németh 2012a, b; Martin and Németh 2006; Sumner et al. 2005). As a consequence of such eruption mechanism, the Harrat Rahat scoria cones are commonly associated with clastogenic lava flows that form upon heat retention of landed pyroclasts that then melt together and form a new melt on the proximal areas of the cones. Harrat Rahat is rich in such volcanic features and that makes the field special in this respect. In some occasions the cones show evidences of lateral spreading as a response to the high heat in their proximal areas and the underlying melt that can function as a lubricant to be able to displace large sectors of cones (Fig. 3.14). Cone rafting is also prominent volcano-morphological features in the field (Fig. 3.15) and it has been commonly accompanied with explosive ash-emitting eruptions that then partially covered the still moving lava flows gradually displaced part of individual cones (Murcia et al. 2013; Riggs and Duffield 2008).

Fig. 3.11 Small and steep lava spatter cone in the volcanic chain of the 1256 AD eruption site in Northern Harrat Rahat [24° 20′ 43.06″N; 39° 46′ 32.13″E]

Fig. 3.12 Large scoria cone with complex stratigraphy potentially represents an eruption site that was active over prolonged time and better to view it as a polygenetic small-volume stratovolcano [24° 21′ 16.13″N; 39° 48′ 49.43″E]

3.3 Volcanic Precinct Concept and Its Benefits

The proposed Harat Al Madinah Volcanic Geopark (HAMVG) (Fig. 3.16) is based on a holistic geoeducation and geoconservation philosophy in order to demonstrate the diversity of volcanism associated with the evolution of long-lived monogenetic volcanic fields in intra-continental regions (Moufti and Németh 2013a). It has been suggested that volcanic features that would form the core of the geoheritage value of the Harrat Rahat should be arranged into a hierarchical system based on the systematic evaluation of each of the geoheritage sites selected on the basis of their value (Moufti and Németh 2013a). Such a system of the volcanic features and landforms preserved in the territory of the proposed HAMVG could emphasize the scientific (geological—volcanological) entity, the level of importance, and the conditions of access to those sites (Moufti and Németh 2013a). This system therefore would be able to offer a self-explanatory guide for end-users to develop alternative

Fig. 3.13 Lava spatter dominated cone with characteristic "collar" in the lip of its crater as a sign of strong welding that preserved the crater rim [24° 11′ 43.74″N; 39° 52′ 51.14″E]

geoeducational programs that are easily linked to specific geological-volcanological topics the designed system can offer (Moufti and Németh 2013a).

Geological (and/or geomorphological) sites have just been started to be catalogued in Saudi Arabia with various level of success and/or detail following the geosite (geomorphosites), geotope and geopark concept that has been successfully used elsewhere (Fuertes-Gutierrez and Fernandez-Martinez 2012; Kazancı 2012; Pulido Fernandez et al. 2014; Vujičić et al. 2011). Recently initiated projects in the Kingdom of Saudi Arabia have identified and documented many volcanic geosites that are significant in their context, such as significant in comparison to the host volcanic region where they are located, as well as carrying value that make them internationally important volcanic features to contribute to the global understanding of specific volcanic processes.

Initially an attempt was pursued to establish the first geopark with a volcanic theme near the culturally important region of Al Madinah city (Moufti and Németh 2013a, b and c). An idea to establish a geopark near Al Madinah was argued on the basis of the high scientific, aesthetic and economic potential the volcanic regions near Al Madinah can carry. A proposal is in consideration currently to evaluate the feasibility to go ahead with focused work to establish such geoparks.

Here we provide the geological and geographical scientific information to provide enough background to show that the region is suitable to develop a volcanic geopark. The scientific research recently intensified on understanding dispersed volcanic systems along the western margin of the Arabian Peninsula that brought a new global interest to explore the volcanic fields abundant in this region (El Difrawy et al. 2013; Murcia et al. 2014; Runge et al. 2014; Wahab et al. 2014; Zobin et al. 2013) many of them was triggered by recent seismic unrest likely been caused by dyke intrusions to a very shallow level of the crust (Baer and Hamiel 2010; Duncan and Al-Amri 2013; Koulakov et al. 2014; Mukhopadhyay et al. 2013; Pallister et al. 2010). The fact that Harrat Rahat also host one of the youngest eruption sites (1256 AD) that are exceptionally well-preserved and located nearby Al Madinah city justify clearly that with the abundance of scientific research that can offer a well-designed volcanic geological model a geopark concept

Fig. 3.14 Collapsing cone of Mosawdah [24° 14′ 13.30″N; 39° 47′ 48.74″E]

can be developed and distinguish these volcanic areas significantly from others on the global scale while can be linked easily to other similar fields elsewhere in the globe along their scientific as well as landscape aesthetic and accessibility value (Moufti and Németh 2013a).

The above outlined logical set naturally offer that in a large area such as any of the harrats in western Saudi Arabia, particularly the Harrat Rahat, the best way to follow some sort of "precinct" concept to link geoheritage sites along their common geoeducational value, and of course their geographical locations (Moufti and Németh 2013a). The "precinct" concept naturally groups together the main and most representative volcanic features (including landforms and associated geotopes) to form at least three distinct precincts as the basis of a proposed volcanic geopark (Moufti and Németh 2013a). The HAMVG's volcanic landforms naturally offer a three-layered precinct hierarchy with an additional extra level which could be linked to more adventure style geotourism as the site located far from the others and can offer a true remote arid region experience to anyone who would visit those locations, in spite that geologically it is not offering anything significantly different than the other precinct (Fig. 3.16). The precinct concept has been applied to geoeducational programs as the core of a geopark concept in other regions, such as the Kanawinka Geopark in southern Australia and Victoria (http://www.kanawinkageopark.org.au/). In comparison to the Kanawinka Geopark's precincts, the proposed HAMVG's precincts are not only thematically but also geographically well-separated, allowing distinct geotourism projects to be designed around them (Moufti and Németh 2013a).

3.4 Volcanic Precincts Versus Volcanic Heritage Routes

Volcanic precinct are favoured against volcanic heritage route design in the case of the Saudi Arabian harrats. Volcanic precincts can offer more than a linear path to explore geoheritage sites along a well-designed route. A precincts can group geosites that are by some reason can be associated with similar geological or geomorphological concept, information or state of research and therefore can be used to target specific audience to visit that sites. In case of the Al Madinah region, the volcanic precincts follow an order that link to the level of adventure tourism needed to explore the

Fig. 3.15 An older rafted scoria cone near the young Al Anahi scoria cone and lava flow field [24° 16′ 38.54″N; 39° 46′ 6.11″E]

grouped geosites with the level of complexity of the volcanological knowledge that could be achieved by a visitor just by completing the specific precinct tours. Near Al Madinah, the 3 + 1 precinct is designed to follow a natural logical path (Fig. 3.16).

Precinct 1 is all about the historic volcanic eruptions that affected the life of the people in the region in the past several centuries, and also had some influence on the cultural development of the region (Fig. 3.16). These sites are easy to access, they are well-preserved, and together they can provide a very good introduction to understanding volcanic processes.

Precinct 2 would involve a little bit longer trip to complete and offers a more detailed understanding of the type of volcanic eruption most common among the harrats, lava spatter cones and extensive lava flows.

In Precinct 3 visitors explore unusual volcano types that formed lava domes, explosion craters, and even produced pyroclastic flows that covered vast areas in the central part of Harrat Rahat, visible on a satellite image (Fig. 3.17). Visiting the selected geosites in Precinct 3 will provide anadventurous geotouristic experience, including evidence to demonstrate the destructive force of explosive volcanism.

Precinct 4 is a more adventurous version of Precinct 3, offered as an alternative for those visitors eager for adventure tourism. The geosites of this precinct are deep inside the interior of Harrat Rahat, and to visit them requires preparation and experience.

Harrat Rahat is appropriate for developing geoeducational and geotouristic projects arranged in precincts rather than in geoheritage routes. Within the precincts, geosites are arranged along routes that link geosites with specific geological value.

3.5 Lava Flow Features and Their Geoheritage Value for Understanding Lava Flow Field Evolution

Harrat Rahat is probably the most accessible harrat in the Kingdom of Saudi Arabia and it is home of a great diversity of lava flow morphotypes (Murcia et al. 2014). The arid climate and the relatively easy access of many of the sites can allow visitors to see most of the lava flow surface textures in their pristine status. The arid climate offers

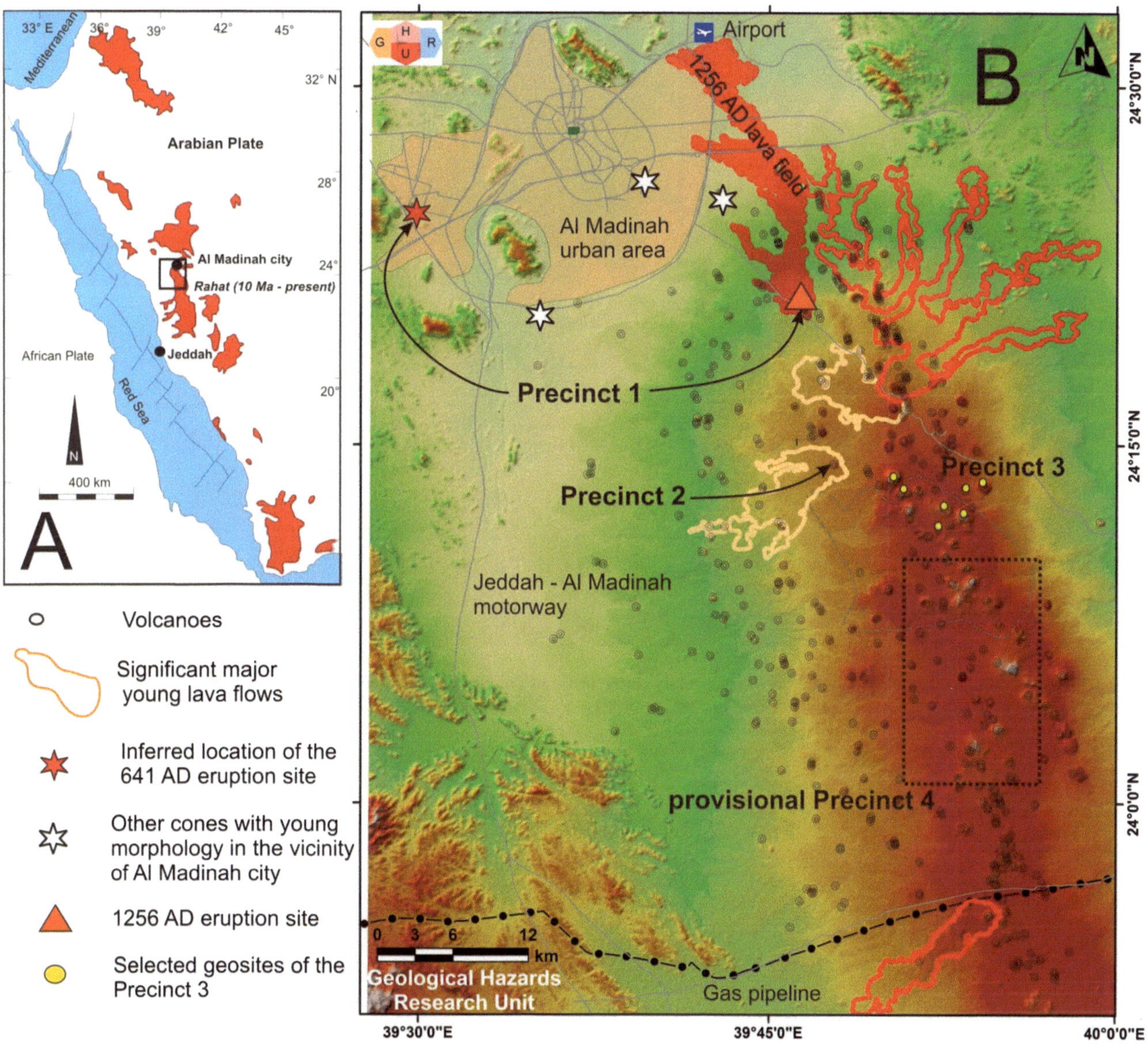

Fig. 3.16 Volcanic "precinct" regions as the core of the provisional volcanic geopark of Al Madinah in the Harrat Rahat

well-preserved lava flow surface textures to be seen. Especially nearby Al Madinah city the extensive road network that are linked to dirt roads across the harrat form an ideal logistic set to select specific lava surface sites to promote and include in various precincts to be listed as key geosites.

The phrase "lava morphotype" refers to the characteristics of the surface morphology of any lava flow after solidification. The lava flow surface morphotypes carry significant information on the cooling history, rheology and the dynamics of the lava flow during its molten stage (Anderson et al. 2012; Duraiswami et al. 2014; Njome et al. 2008; Solana 2012; Suh et al. 2011; Woodcock and Harris 2006). In the Kingdom of Saudi Arabia, young and well-preserved mafic lava fields display a wide range of these morphotypes (Murcia et al. 2014).

At Harrat Rahat a framework of lava surface morphotypes for describing changes in morphotypes down-flow has been proposed (Murcia et al. 2014). The gradual changes of lava surface morphotypes can be traced very clearly along the 23 km long 1256 AD historic lava flows (Fig. 3.18). The changes over distance provide an important scaling aspect for the visitor to appreciate the lava flow emplacement mechanism. The abundance of small-volume and short lava flows can also be used to develop geoeducational projects to demonstrate the variability and the unique nature of a silicate melt to flow on the Earth surface.

Implications of demonstrating the variety of lava flow surface textures through a well-designed geoeducation program can contribute significantly into the volcanic hazard education of the local population. The lava fields of the

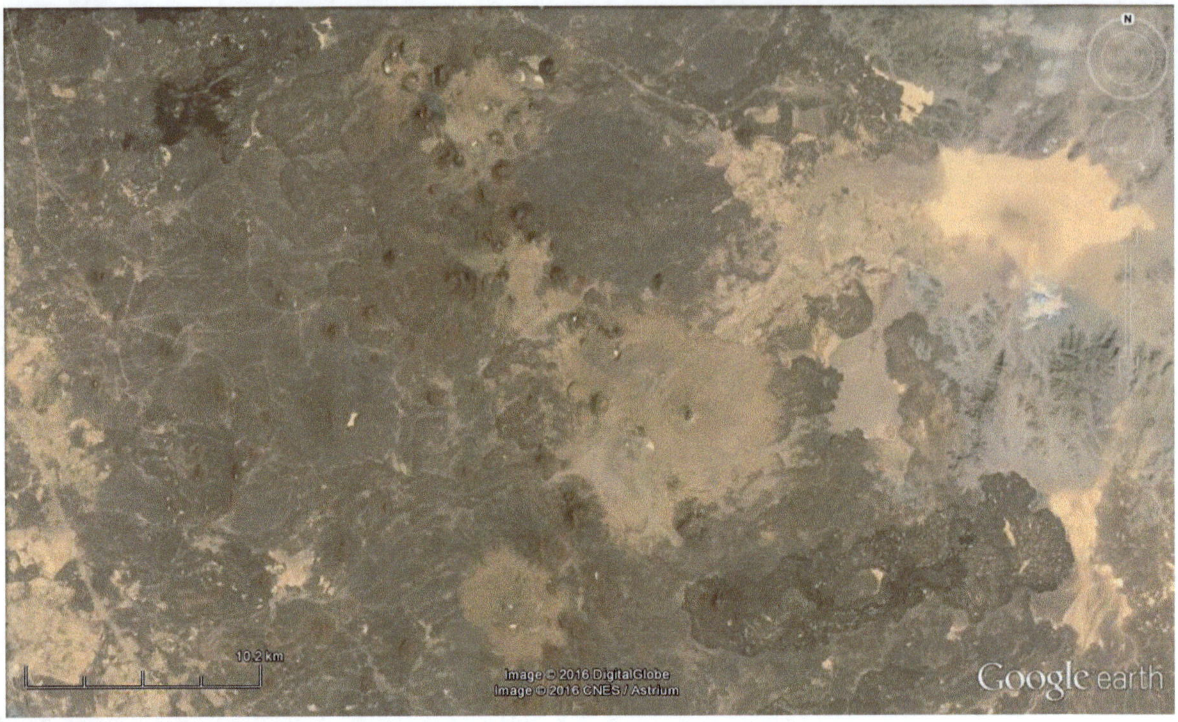

Fig. 3.17 Satellite image in the Precinct 3 showing visually different (light coloured) regions in the central part of Harrat Rahat. The top left corner of the map view is about [24° 14′ 46.07″N; 39° 39′ 48.11″E]

Fig. 3.18 Lava surface morphotype change across the 1256 AD lava flow main axis [24° 23′ 2.38″N; 39° 46′ 13.77″E]

recently proposed Al-Madinah Volcanic Geopark, and other harrats are commonly mentioned as continental flood basalts or large igneous provinces (White et al. 2009) and they carry important geoeducational value (Fig. 3.19).

Overall, the Harrat Rahat lava flow fields extend up to 23 km from the source, and vary between 1–2 and 12 m in lava flow thickness (Murcia et al. 2014). The lava flow fields cover areas between $\sim$32 and $\sim$61 km^2, with individual volumes estimated between $\sim$0.085 and $\sim$0.29 km^3 (Murcia et al. 2014). The lava flow surface textures exhibit shelly-, slabby-, and rubbly-pahoehoe, platy-, cauliflower-, and rubbly-a'a, and blocky morphotypes roughly in this order to downflow (Murcia et al. 2014). The specific lava flow surface textures are linked to both intrinsic (i.e. composition, temperature, crystallinity and volatile-content/vesicularity) and extrinsic (i.e. emission mechanism, effusion rate, topography and flow velocity) emplacement parameters and their changes over distances (Murcia et al. 2014). In many places along the 1256 AD lava flow one morphotype transitions to another in individual flow-units or lobes and that they dominate zones (Murcia et al. 2014).

Pahoehoe morphotypes (Fig. 3.20) are more related to the simple mechanical disaggregation of the solidified crust over an inflating lava flow body that under mechanical stress gradually disaggregating and carried away by the moving mass (Murcia et al. 2014). A'a morphotypes in the contrary are related to the transitional emergence and posterior fading of clinker, and blocky morphotype to fracturing and auto-brecciation. a'a morphotypes (i.e. platy-, cauliflower-, rubbly-a'a) are those that dominate the lava flow field surfaces in northern Harrat Rahat, which suggests that core-dominated flows were predominant during flow movement (Murcia et al. 2014). Lava structures may be related to some morphotypes once they were well-developed and they are well-preserved. In particular, down-flow changes exhibit key illustrative and easy recognizable features in the lava flow fields and might provide insights into real-time monitoring of future flows in this region. From geoeducational point of view Harrat Rahat offers probably the most accessible and the greatest diversity of lava flow types, and therefore these flow fields can help to show evidence and consequences of potential lava flow eruption event in the future.

Fig. 3.19 Lava surface texture changes in proximal areas through cascading lava flows (dark centre of the flow) over steep slopes [24° 21′ 5.00″N; 39° 46′ 31.09″E]

Fig. 3.20 Pahoehoe lava surface morpfotype [24° 21′ 8.66″N; 39° 46′ 32.74″E]

3.6 Volcanic Cones and Their Geoheritage Value

Volcanic cones are abundant in the territory of Harrat Rahat and they range from a very small ($\sim$10 m high) to cones that are over 100 m above their surroundings. Cone morphology reflects some degree of their age, and potentially could be used for relative age datings such as it has been suggested and trialled elsewhere (Porter 1972; Settle 1979; Wood 1980a, b). Relative age dating of the volcanic cones based on cone morphometry is, however, in arid climate might not work in a way how early studies predicted as the cone geometry modification is a very slow process and cones can appear in a very youthful appearance after significant time as demonstrated in many scoria cone fields (Kereszturi and Németh 2012a, b).

Erosion of scoria cones show a great variety of trends that are commonly linked to relative age (Doniz-Paez 2015; Fornaciai et al. 2012; Inbar et al. 2011; Inbar and Risso 2001; Kervyn et al. 2012) (Porter 1972; Settle 1979; Wood 1980a, b; Hooper and Sheridan 1998; Doniz et al. 2008; Inbar et al. 2011; Fornaciai et al. 2012), but recent studies also show that especially in cases when cones are dominated by lava fountain-fed spatter eruptions, the cone erosion, and cone geometry-modification more commonly linked to the cone genetic history or the substrate morphology than to the time passed since its eruption (Favalli et al. 2009; Kereszturi et al. 2013; Kereszturi and Németh 2012a, b). While scoria cones in Harrat Rahat show a great variety of erosional stages, e.g. common gully formation on their flank, the majority of the cones are easy to recognize and they are spectacular (Fig. 3.21). Young cones are relatively smooth surfaced and their craters have no aeolian dust infill. Their craters are rarely breached however crater breaching across the central part of the Harrat Rahat is common as response to the edifice instability caused by the extensive lava flow outpouring in their flank region.

While the youngest scoria cones are easy to recognize in the field and in satellite imagery, to use their morphometry parameters for relative age dating can be misleading and likely cannot provide high resolution of ages to be able to distinguish even a Pleistocene cone from a Holocene one. An excellent example is the 641 AD four cones just SW of Al Madinah city. The four cones are only inferred to be the eruption sites of the 641 AD historical eruption however their morphology cannot be distinguished from other

Fig. 3.21 Typical, moderately erosion modified large scoria cone in the Harrat Rahat [24° 23′ 0.45″N; 39° 49′ 59.20″E]

Pleistocene to Holocene cones. This problem is partially due to the fact that this cones are dominated by lava spatter eruption, similar to many other cones in the Harrat Rahat, that formed collar-like spatter ramparts in their crater rim, that acted as a preventing shield in top of the cones, lowering significantly the erosion speed, and change the style of erosion as predicted in recent studies (Kereszturi and Németh 2012a, b).

3.7 Lava Domes and Explosion Craters as the Results of the Potentially Most Hazardous Volcanism in the Region

Results of silicic volcanism are clearly visible as spectacular volcanic landforms in three of the harrats in Saudi Arabia; in Harrat Rahat (Moufti and Németh 2013a), Harrat Khayber (Moufti and Nemeth 2014) and Harrat Kishb (Moufti et al. 2013c). Harrat Rahat however is the location where these volcanic features are relatively young, about 0.3 to 0.7 Ma (Camp and Roobol 1989; Moufti et al. 2012), therefore well-preserved and in addition they can be accessed through dirt roads by a relatively short highway drive out of Al Madinah city. The centrally located position of these silicic volcanoes can provide a great opportunity to both adventure and eco-tourism as the sites are remote and commonly provide unique landscape forms with unique ecosystems (Fig. 3.22).

The central part of the Harrat Rahat is covered by various silicic (mostly trachytic) pyroclastic deposits forming an extensive ash plain landscape that are surrounded by steep and high lava domes. The lava domes are interestingly commonly associated with older scoria cones and it seems they erupted through pre-existing volcanic landforms. Their composition ranges from mugearite through benmoreite to trachyte (Camp and Roobol 1989).

The volcanic landforms are diverse, and can be seen simple lava domes, lava dome complexes, and lava dome complexes associated with explosion craters commonly deep as over 100 m (Fig. 3.23). The presence of the silicic eruptive centers carries a significant volcanic hazard aspects putting the Harrat Rahat among those volcanic regions where violent explosive eruptions and associated lava dome formation was far more common as it is considered, and in spite of the relative older age of these sites, such future

Fig. 3.22 Matan lava dome complex is a unique geotop [24° 13′ 26.43″N; 39° 50′ 22.23″E]

Fig. 3.23 Typical explosion crater of the Gura 2 in the central region of the Harrat Rahat [24° 12′ 22.95″N; 24° 12′ 22.95″N]

eruptions cannot be excluded. In this respect to develop a geoeducational program around the silicic eruption centers of Harrat Rahat bears a very important aspect of the proposed Al Madinah Volcanic Geopark design. The sites are not only exotic, exciting and visually outstanding, but also offer a different angle to demonstrate the volcanic eruption styles formed the landscape at Harrat Rahat.

From the scientific perspective, the common presence of small to medium volume silicic lava domes in a volcanic field suggests that some degree of crustal storage network must exist beneath the Harrat Rahat to form chemically evolved magmas in spite of the general dispersed, volcanic field-forming nature of the volcanism. The relatively small-volume and simple architecture of the lava domes of

Harrat Rahat makes them different from those lava domes commonly associated with major central (composite and strato-volcano or caldera) volcanoes such as Merapi, Indonesia (Abdurachman et al. 2000), Unzen, Japan (Fujii and Nakada 1999) or Soufriere Hills in Montserrat (Bourdier and Abdurachman 2001; Carn et al. 2004). The lava domes of Harrat Rahat are single, individual sites that were probably grown over decades, but their eruption was likely controlled by a single or low number of eruptive phases that make them closer relationship with typical monogenetic volcanoes than to those complex and long-lived lava domes commonly developed on top of major long-lived polygenetic volcanoes. In this respect, Harrat Rahat's lava domes can offer a unique opportunity for scientific research to understand how dispersed lava dome field evolve, and how they contribute to the geological record of volcanic fields.

The silicic lava domes of Harrat Rahat are similar to those lava dome fields documented in association with dispersed small-volume volcanic fields such as those in Central Mexico (Blatter et al. 2001; Guilbaud et al. 2012; Hasenaka 1994; Hasenaka and Carmichael 1985; Riggs and Carrasco-Nunez 2004) or in SW US (Riggs et al. 1997). Similar dispersed lava dome fields have been documented across the Miocene to Pleistocene Carpathian Volcanic Arc (Lexa et al. 2010) that highlight the significance of such small volume lava dome systems in regard to understanding their origin as part of a dispersed volcanic region or volcanic field. In scientific perspective the lava domes of Harrat Rahat are significant features, and can offer key sites to study lava dome formation, their geomorphological evolution and their effect on the surrounding regions through block-and-ash flow eruptions.

In addition to lava domes the Harrat Rahat also host numerous explosion craters (Fig. 3.23). These craters are diverse in their size (crater diameter and depth) and exclusively located in the central part of the field. In the crater wall of these craters commonly half section of older silicic lava domes are exposed indicating some link between lava dome growth and sudden disruption and crater formation. The smallest craters are surrounded by coarse pyroclastic breccias inferred to be explosion breccias. These deposits are rich in accidental lithic fragments and deep crustal origin xenoliths. The juvenile pyroclast content of these pyroclastic rocks are relatively low. The juvenile pyroclasts are low vesicular microlite-rich rocks indicating potential magma-water explosive eruptions as a cause of their fragmentation.

In larger craters such evidence to support potential magma and water explosive interaction is less clear, and the deposits surrounding the craters are more typical block-and-ash flow deposits typical for moderate run-out distance pyroclastic flows. In this respect Harrat Rahat's explosion craters can be classified as small maar volcanoes to more typical broad craters with even moderate caldera collapse features in their summit. The important aspect of these explosion craters beside the gradual trend from phreatomagmatic to magmatic explosive types of eruption as a driving force to their formation is, that even the largest and most complex craters are relatively simple in comparison to a long-lived silicic volcano. This fact again offer a unique scientific background to establish a scientifically well-established geoeducation program to demonstrate the full spectrum of eruption styles and volcano types associated with a predominantly dispersed, volcanic field building volcanic system such as Harrat Rahat.

3.8 Organisation of Precincts of the Proposed Harrat Al Madinah Volcanic Geopark (HAMVG)

The HAMVG three precincts (Fig. 3.16) are proposed as based on their scientific aspects, level of exposures, geoeducational value and logistical aspects (Moufti and Németh 2013a):

Precinct 1 Historic Eruption Precinct—1256 AD and 641 AD Historic Eruption Sites;

Precinct 2 Lava Lakes, Lava Fountains and Volcano Spreading Precinct—The Mosawdah Volcano and

Precinct 3 From Silicic Lava Domes to Explosion Craters Precinct.

Precinct 4 An additional Precinct has been outlined as an alternative or extension of the Precinct 3 to provide a stronger adventure touristic aspect to fundamentally the same geological processes Precinct 3 can demonstrate.

Precinct 1 groups volcanic features and associated geoeducational and geotourism programs that demonstrate the eruption sites that have been historically documented and are probably the most relevant to the inhabitants of Al Madinha city. The key to establish this precinct is the direct relevance of the demonstrated volcanic features to the life of the locals. This precinct hosts numerous geosites dealing with extensive transitional pahoehoe-to-aa lava fields with world-class examples of lava flow surface textures, lava spatter and scoria cone (Murcia et al. 2014).

Precinct 2 can be viewed as an expansion of the first, offering the visitor a more in-depth understanding of the type of volcanism very common in the Harrat Al Madinah. Precinct 2 is centered around the main volcanic geotope, the Mosawdah Volcano and its lava flows, and its numerous geosites that provide superb examples to understand the life of a highly active effusive volcano that formed lava fountaining. As a result the volcano under its own erupted hot material gradually collapsed and signs of the edifice spreading and collapse are evident. It is more difficult to access the geosites of Mosawdah volcano than the geosites of Precinct 1 and therefore it would involve some "adventure tourism" style trip which makes this precinct available only to those visitors who wish to go deeper into understanding volcanic processes. Precinct 1 and 2 fundamentally cover the majority of the volcanic features that can be located in the Al Madinah Volcanic Field (Moufti and Németh 2013a).

Precinct 3 offers the most adventurous trips for visitors and some unique additions to understanding the full spectrum of volcanic processes in the AMVF. Precinct 3 volcanic features deal with silica-rich volcanism that formed various lava domes (e.g. trachytic), as well as deep explosion craters, many of them at least in their initial phase were formed due to magma and ground-water explosive interaction. Precinct 3 is located far from Al Madinah city, and only well-equipped geotourists with trained guides can visit the sites. While the volcanic features in Precinct 3 can be seen to have a very high aesthetic and scientific value, they are rather an extra addition to the full picture of the volcanism of the AMVF, than something without which the visitor would get a distorted image of the field. However, those who decide to invest energy to visit Precinct 3 would be well rewarded by a truly dramatic volcanic landscape. Precinct 3 could be expanded toward the south (**provisional Precinct 4**) as an alternative geoheritage site, where a great variety of pyroclastic flow deposits and associated volcanic craters can be visited. These sites have a very unique landscape value. However, visits to these sites can only be done by well-prepared adventure tours.

3.9 Precinct 1—Historic Volcanic Eruption Sites

The largest historic eruption at 1256-AD that lasted ~52 days produced about minimum 0.29 km^3 lava forming a maximum of ~23-km long and 8-m thick flow field (Murcia et al. 2014). This complex semi-confined to unconfined lava field is dominated by transitional flow textures typical for fast moving, open channel lava. Gradual transition from a shelly-, slabby-, and rubbly-pahoehoe, toward platy-, cauliflower-, and rubbly-a'a, reflects lava flow rheology changes (Murcia et al. 2014). The resimulation of the 1256-AD flow with MAGFLOW code (Bilotta et al. 2012; Cappello et al. 2011; Del Negro et al. 2008; Herault et al. 2009) suggests also a complex flow evolution, including late stage ponding of lava around the emission points (Nemeth et al. 2013). Flow inflation/deflation features such as lava rises, tumuli, lava blisters, pressure ridges and evidences of cone rafting are common in proximal areas (Camp et al. 1987; Nemeth et al. 2013). The cones of the 1256-AD eruption are dominated by flattened lava spatter, ash, lapilli with Pelee's hair and tears, and reticulate suggesting lava fountain-dominated eruptions as well as Strombolian style explosive eruptions (Murcia et al. 2013). Textural features are common for lava lake level fluctuations and lava outbreaks inferred to cause edifice spreading (Nemeth et al. 2013). Other historic eruption took place in 641-AD forming four small cones—recently named as Al-Du'aythah volcanic cones (Murcia et al. 2015)—aligned in NNW–SSE in the western edge of Al-Madinah City (Moufti et al. 2013b). Three out of the four cones has basal phreatomagmatic deposits indicating initial phreatomagmatic explosions (Moufti et al. 2013b; Murcia et al. 2015). This is the only location in the younger (<10,000 years) eruptive centers in the northern Harrat Rahat where evidences of phreatomagmatism are known (Murcia et al. 2015). The 641-AD cones' upper sequences inferred to be produced by typical lava fountain and moderate Strombolian style explosive eruptions that produced small clastogenic flows reaching less than 300 m from their source (Murcia et al. 2015).

While lava spatter and scoria cones are among the most common volcanic landforms on Earth (Németh 2010; Valentine and Gregg 2008; Vespermann and Schmincke 2000), to see perfectly exposed and unmodified landforms is becoming increasingly difficult because they are either remotely located, have suffered from significant anthropogenic modifications or they are in areas where the vegetation cover inhibits views of the original landscapes. The Precinct 1 "*Historic Eruption Precinct—1256 AD and 641 AD Historic Eruption Sites precinct*" of the proposed HAMVG comprises volcanic landforms that are well exposed, easy to access and record a unique volcanic process associated with a sustained fissure-fed volcanic eruption, considered to be one of the last major volcanic eruption in the Arabian Peninsula (not counting on those volcanoes erupted recently in the axis of the Red Sea (Xu and Jonsson 2014). Volcanic phenomena represented in this precinct include the results of prolonged lava fountain fed eruptions, such as cone rafting and associated lava lake infill and drain-back, as well as lava flow outbreaks at various points on the fissure-axis edges of the developed volcanic cones. The variety of volcanic features associated with lava fountain type volcanic eruptions is great and ranges from identification of traces of lava-lake level fluctuations in the inner crater walls and clastogenic (rootless) lava flow formation through rapid accumulation of lava spatter in the inner and proximal outer flank of the volcanic cones, to rock records that document fully developed and established volcanic conduit conditions promoted by Strombolian style magmatic gas bubble outburst-driven explosive dispersal of pyroclasts, forming extensive tephra blankets (Hintz and Valentine 2012; Keating et al. 2008; Valentine 2012; Valentine and Gregg 2008; Valentine et al. 2007).

The 1256 AD eruption site with its complex cones along a 2.3 km long fissure form a complete volcanic geotope (Moufti and Németh 2013a; Moufti et al. 2013d and e; Murcia et al. 2013), therefore, it is a significant educational site, where visitors can learn about the complexity of magmatic effusive and explosive eruption styles that may occur along a long lived and evolving fissure, the interaction between effusive and explosive stages of eruptions, as well as the link between changes of eruptive rate and the resulting volcanic landform (and landscape), and the dynamic processes that may take place in volcanic craters. The proximity and easy access to the Precinct 1 "*Historic Eruption Precinct —1256 AD and 641 AD Historic Eruption Sites*" to the Al Madinah city, coupled with the young age of the eruptions and the historical documentation, make this site the perfect location to provide eye-opening evidence of the style of eruptions the region may face in the future.

3.9.1 Geotope of the 1256 AD Historic Eruption Site and Its Lava Flows

The 1256 AD eruption produced Hawaiian and Strombolian style volcanic activity through a ~2.2 km long fissure that created seven individual and nested volcanic cones (Fig. 3.24). The individual cones are aligned along the fissure and some are partially destroyed. The smallest cone is less than 0.1 km wide and 10 m high, while the biggest cone is 0.7 km wide and 90 m high. The smallest cones are typical lava spatter cones with steep flanks and agglutinated and welded banks of spatter and clastogenic flow lobes (Moufti et al. 2013e).

The 1256 AD eruption site is a perfect geotope in the sense of its geological heritage. It is composed of individual volcanic cones erupted in similar style and produced overlapping pyroclastic rock units as well as multiple lava flows. The link between individual geosites are along the fact that they have been produced by similar geological processes slightly differs from each other as the controlling parameters for each eruption was a little bit different. As a result, it is very clear to define the boundary of the geological feature (as the cones along the fissure), easy to link them together along a common geological process (the Hawaiian to Strombolian style eruptions), they are easy to distinguish from other parts of the volcanic field, e.g. they form the volcanic edifices and their proximal areas, that are different geologically then the inter-cone regions where extensive ash plain formed from pyroclastic falls (tephras).

This distinction and separation of these volcanic landforms from others also logical and scientifically valid as they follow the boundary between the volcanic edifice and the surrounding volcanic ring plains and their deposits as it has been outlined in many other areas and many other type of volcanoes (Kereszturi and Németh 2012a, b; Manville et al. 2009; White 1989, 1990, 1991). The 1256 AD eruption site as a geotope offers a great variety of geoeducational sites to be presented. The individual geosites were identified on the basis of their scientific information they can provide, their preservation potential and attractiveness for both professional and general audience.

Pyroclasts that were ejected beyond the 1256 AD cone's (medial-to-distal pyroclastic succession), forming a tephra cover composed of angular-to-plastically shaped pyroclasts including Pele's tears, hair and basaltic reticulate (Kawabata et al. 2015; Nemeth et al. 2013). Recent study of the distribution pattern of the ash plain around the cones reviled that the eruptions that produced the pyroclastic fall fed from multiple eruption plumes each representing individual eruption phases (Kawabata et al. 2015). In proximal areas such tephra sections are particularly well-exposed and can offer great geosites to define where visitors can see that even predominantly effusive and mild explosive eruption-dominated volcanoes such as the 1256 AD eruption sites can be associated with extensive tephra fall-producing eruptions. Such researches are recently been conducted in other places on Earth (Németh et al. 2011; Valentine and Gregg 2008; Valentine and Keating 2007; van Otterloo et al. 2013) and therefore the 1256 AD eruption site geotope can be easily linked to those front-line researches and can offer a new aspect to understanding mafic explosive eruption processes and their volcanic hazard aspects.

Besides the volcanic processes that built the cone and "ash-plain", cone rafting, as well as central crater floor subsidence, took place frequently, leaving behind truncated cones and pyroclastic raft-covered lava flows to complicate the volcanic landforms. Medial-to-distal ash and lapilli were accumulated on and below lava flows, indicating coeval lava effusion, pyroclastic fall producing explosive eruptions and lava flow outbreaks in cone-flank regions. Horseshoe-shaped cones, multiple rafted cones and nested cones create a diverse and complex volcanic landscape along the fissure. Each of the volcanic cones shows evidence of crater floor subsidence, crater wall collapse (Fig. 3.25), and some degree of rafting of the cone's outer flank through lava outbreaks in the foothill of the cones, suggesting that the craters of these cones were filled with lava lakes with fluctuating levels. The lava spatter-dominated proximal volcanic successions are interpreted to be deposited from Hawaiian style lava fountaining and Strombolian-style explosive eruptions. Lava spatters erupted through lava fountains accumulated along the vents, where the freshly deposited hot material formed agglutinate that commonly fed clastogenic lava flows (Fig. 3.26). The heat from the lava lakes and from the fast accumulated lava spatter piles favoured the perfect physical conditions for the freshly-landed pyroclasts to coalesce and form increasingly steepening cone morphologies. Slope

Fig. 3.24 Overview of the 1256 AD eruption site looking from the NW

angles are commonly higher than the angle of repose of any granular material, due to the agglutination and coalescence of individual pyroclasts on the flank of the growing lava spatter cone.

The complex semi-confined to unconfined lava fields of the 1256 AD eruption are dominated by transitional flow textures typical for fast moving, open channel lava (Bretar et al. 2013; Duraiswami et al. 2003, 2014; Wantim et al. 2011). The gradual transition from a shelly-, slabby-, and rubbly-pahoehoe, toward platy-, cauliflower-, and rubbly-a'a, reflects lava flow rheology changes (Murcia et al. 2014). Flow inflation and deflation features, such as lava rises, tumuli, lava blisters, pressure ridges and evidence of cone rafting, are common in proximal areas. Textural features of solidified lavas in crater settings are common for supporting repeated lava lake level fluctuations and lava outbreaks inferred to cause edifice spreading.

In the following section a summary of individual identified geosites are listed with a short description. The selection was conducted by a scientific evaluation of the sites ranking their scientific importance, uniqueness and their location. While similar features selected and listed below are abundant in the Harrat Rahat, the selected geosites are those that can be easily accessed and/or linked to a broader educational program including the previously introduced precinct concept.

3.9.1.1 Geosite 1—Southern Cone and Hornito Field [24° 20′ 28.37″N; 39° 46′ 39.97″E]

This geosite is located in the southern margin of the 1256 AD fissure aligned cone chain (Fig. 3.27). It is slightly offset from the main cone edifice which has been partially destroyed by lava flow rafting and the engulfment of flow lobes initiated from the interior part of the crater of the southernmost cone of the 1256 AD eruption site. The southern cone itself is a partially destroyed volcanic landform. Its crater still preserved but its outline difficult to trace due to the thick lava flow covers that truncated its margin. It is relatively well preserved in its eastern side, where it is the highest. In the highest point a large hornito can be seen with a deep cavity as a pipe along magma was squeezed out forming lava spatter-dominated ramparts. In the southern side of the cone, in the outer flank of the edifice a chain of hornitos form a spectacular set of volcanic landforms. Each of the hornitos is around 5 m tall and few metres wide. The

Fig. 3.25 Evidences of crater wall collapse along the crater wall of the 1256 AD cones [24° 21′ 7.64″N; 39° 46′ 21.26″E]

lava spatter forms a chain of steep and narrow cones that has a large cavity supported by the agglutinate wall of the hornito (Fig. 3.28). The significance of this site is to demonstrate the fact that lava lake level fluctuations took place along the 1256 AD fissure eruption, and sometimes lava was squeezed out along marginal structures forming chain of hornitos. While similar geosites have been named elsewhere (Gao et al. 2010, 2013) and such processes can be observed in their stage of formation in Hawaii, this site is significant due to its easy access, large geometry and important message to understand late stage volcanic hazards such an eruption site can produce (Ort et al. 2008).

3.9.1.2 Geosite 2—Southern Cone and Lava Tube Field [24° 20′ 37.01″N; 39° 46′ 37.12″E]

This geosite has been defined in the NE section of the outer edifice margin of the southern cone of the 1256 AD eruption site geotope (Fig. 3.27). In this site thin crusted lava flows forming a complex network of lava tubes (Fig. 3.29). The lava tube are clearly inflational features of low viscosity and fluidal, but gas rich lava as suggested by the high vesicularity of the tube-forming rocks. The tubes are about a meter wide, box-shaped in cross sectional view and surrounded by about 5 to 20 cm thick porous (vesicular) chilled crust. The interior of the tubes are rich in small-scale (cm-scale) lava stalactite and stalagmite. The individual lava tubes are cross-cutting each other and seem to form a positive relief on the landscape indicating that the low viscosity melt outpoured and quickly run down on the emitting cone flan. The significance of this geosite is to demonstrate the effect of low viscosity of lava on the resulting lava tubes. Similar features are common elsewhere in the Saudi harrats, but this site easy to access nature makes this location a valuable geosite. This geosite also highlight the need of nature conservation projects in the region, as these sites are fragile and easy to be demolished.

3.9.1.3 Geosite 3—Steep Lava Spatter Cones [24° 20′ 42.23″N; 39° 46′ 32.01″E]

This geosite [24° 20′ 45.53″N; 39° 46′ 30.55″E] is the third distinct cone (Fig. 3.27) of the 1256 AD Al Madinah Volcanic Eruption Geotope. This geosite is located between cone

Fig. 3.26 Clastogenic lava flows in the edifice building succession of the 1256 AD eruption site [24° 21′ 22.18″N; 39° 46′ 15.53″E]

2 and 4 (Fig. 3.27) and its morphology differs significantly from the morphology of the cone 2 and 4 (Fig. 3.30). It has a very steep slope angle, being almost 60° along the crater rim and over 35° at its base that makes this cone among the steepest known cones on Earth (Moufti et al. 2013e). These slope angles are higher than the angle of repose of any granular material, and account for the agglutination and coalescence of individual pyroclasts on the flank of the growing lava spatter cone to maintain these slope angle value (Moufti et al. 2013e). The cone has a crater about 20 m deep with vertical wall that is draped by lava spatter suggesting some drainage and refill of the crater by repeated lava injection. While steep lava spatter cones are common features on Earth, this site is unique as this cone probably would fit to the largest of such cones on Earth. In addition its architecture closely resembling a large hornito and raises an important question how it was fed. Was it directly connected to a feeder dyke below, or was it just fed from sideway, as a response to the changes of the physical conditions of the lava lakes hosted in the nearby large cones? The existence of such questions ensure that the location indeed can provide valuable information to our common understanding of lava spatter cone formation particularly to their plumbing system. Such questions and aspects of a geosite can just make stronger why this location been selected as a geosite (Fig. 3.31).

3.9.1.4 Geosite 4—Ponded Pahoehoe Proximal Lava Fields and Lava Caves [24° 20′ 48.06″N; 39° 46′ 16.35″E]

In the western side of the main central cone of the 1256 AD eruption side geotope is a proximal area of lava flows erupted from the 1256 AD fissure (Fig. 3.27). This location is one of the most diverse and easy to access in the entire Harrat Rahat in respect of the variety of lava ponding types one can visit in a relatively small area. Its diversity is great in regard of the variety of lava surface textures the visitor can explore as well as the numerous evidences of inflation and deflation of ponded lava flows. The site is an easy walk distance from a sealed road from where with a less than an hour walk the visitor can explore the effect of lava ponding and formation of various features define significant time (days to weeks) when magma was just ponded in the depressions surrounded the growing cones.

The area is best defined as a large silver, grey smooth surfaced region where hummocky surface of the lava flow

Fig. 3.27 Overview map of the location of identified geosites associated with the 1256 AD eruptions site's geotope. Geosites numbered as listed in the text. Note that the *upper* image shows the northern distal lava flow sites while the *bottom* image shows the geosites identified in the proximal area of the 1256 AD eruption site. Cone numbers are also shown on the *bottom* image

Fig. 3.28 Chain of hornitos in the southern edge of the southernmost cone of the 1256 AD eruption site [24° 20′ 27.83″N; 39° 46′ 40.27″E]

Fig. 3.29 Lava tubes forming a complex network of small and narrow lava tubes in the NE sector of the southernmost cone of the 1256 AD eruption site geotope [24° 20′ 36.70″N; 39° 46′ 36.75″E]

Fig. 3.30 Steep lava spatter cone (*right*) as a geosite along the fissure aligned cones of the 1256 AD eruption site geotope [24° 20′ 44.74″N; 39° 46′ 30.72″E]

can be observed (Fig. 3.31). The lava surfaces are smooth with some pahoehoe surface texture marks. Large blocks of smoothed surfaced lava flow fragments are separated by fractures along some dm-scale displacements are common, where the internal texture of the lava crust can be studied. The lava crusts are normally in a dm-scale in their thickness but nearly 1 m thick crusts are also known in the interior of this field indicating that lava ponding must have been taking place over several days or weeks (Holcomb 1981; Polacci and Papale 1997). This observation fits well to the known longetivity of the 1256 AD eruption and this geosite can provide some graphic insight how such historic account could be justified by pure geological observations, which made this geosite an important addition to the

Fig. 3.31 Hummocky surface of ponded lava flow region west of the main cones of the 1256 AD eruption side geotope [24° 20′ 44.50″N; 39° 46′ 18.70″E]

Fig. 3.32 Large lava tube exposed just west of the main cone of the 1256 AD eruption site geotope [24° 20′ 49.00″N; 39° 46′ 16.05″E]

geoeducational programs the Harrat Al Madinah Volcanic Geopark could provide.

This geosite also show fantastic examples of lava inflations and deflations in the form of lava tubes (Fig. 3.32). One of the largest and longest lava tubes has been found in Harrat Al Madinah area. Complex lava tube network as a main artery of the proximal feeding system of the 1256 AD eruption main lava flow is suspected to be located in this area (Murcia et al. 2014). The geosite is a unique and an easy to access location where the visitors can get an immediate insight how the proximal feeding system of major lava flows can function. This site has a very high educational value as it provides evidence that magma can stay hot and fluid long time beneath the growing and thick crust. Such information

Fig. 3.33 Tumuli on the surface of the smooth pahoehoe lava flow field just west of the main cone of the 1256 AD eruption site geotope [24° 20′ 45.24″N; 39° 46′ 15.63″E]

is very important to convey as it carries important volcanic hazard aspects to the local population.

The evidences of lava ponding, inflation and deflation also provide better understanding how large volume of lava be able to accumulate prior it can break out and feed long lava flows. The evidences of the inflation and deflation in this geosites are the large tumuli (Fig. 3.33), collapsed and displaced lava tube roof blocks and abundant occurrences of lava marks in the tubes. This geosite potentially could function as a main site to educate the public about volcanic hazards.

3.9.1.5 Geosite 5—Pressure Ridges, Flow Channels and Convection Zones [24° 21′ 4.74″N; 39° 45′ 46.32″E]

This geosite is located right next to a sealed highway, and therefore it is very easy to access (Fig. 3.27). It shows lava surface features indicating for the lava flow field dynamics and mechanical properties. The elongated, "pathway-like", slightly twisted zone shows that lava flow must have been fairly viscous to form some squeezed zones in what the still molten lava formed draping features as well as the degassing formed a highly vesicular but chilled outer margin of the flow (Fig. 3.34). An almost 50 m-long spreading ridge with a central crack is exposed and ready to be examined by the visitors that form the core of this geosite (Fig. 3.34). Such spreading ridge indicates that the lava crust has formed by lateral spreading from a convective plume similar to other open channel transitional flows such as those located and documented from Krafla in Iceland (Rossi 1997). The rugged vesicular surface of the lava was fed from the crack in the middle as an up-flow of the cooling viscous melt along the ridge. The asymmetry of the features (in map view) can be associated with the differential shear acted upon the entire lava flow field. Along this location abundant evidences can be observed to see, that the lava flow field was time to time fed by newly arrived open-channel fed melt that cut through and just gradually mingled with the ponded interior of the lava. This location provides insight into the dynamic nature of a large ponded lava zone in the proximal areas of a volcano. In addition this geosite contributes to our understanding of the unstable nature of a ponded lava body that can collapse and cause catastrophic flow inundation downhill from the crater.

Fig. 3.34 Rotational feature as a pressure ridge in just west of the main cone of the 1256 AD eruption site geotope suggesting dynamic picture of the lava flow movement in the proximal areas [24° 21′ 4.05″N; 39° 45′ 45.99″E]

3.9.1.6 Geosite 6—Reticulite Field [24° 20′ 51.22″N; 39° 46′ 17.88″E]

In the western foothill of the central cone of the 1256 AD eruption site geotope red scoria with dark glassy lava spatter bomb fields together provide the location for this geosite (Fig. 3.27). This location is part of the transition of the cone edifice and the surrounding ash plain. The presence of reticulite (Mangan and Cashman 1996; Powers 1916) in this site is an important indicator that this eruption erupted very low viscosity magma that was able to produce highly vesicular glassy pyroclasts that are very light and ready to be carried away for long distances (Fig. 3.35). This physical property of a pyroclast is important to constrain the potential that such eruption is capable to produce eruption cloud that can be carried away far and produce ash plain such as in the case of the 1256 AD eruption. The presence of reticulite and Pelee's tears and hair also indicates that the magma fragmentation was largely controlled by sheer of the low viscosity melt upon exiting the crater, and likely to be associated with a lava fountain fed eruption (Mangan and Cashman 1996). This geosite therefore plays an important role to explain the explosive phase of the 1256 AD eruption and carries key volcanic hazard aspect the local community could learn here.

3.9.1.7 Geosite 7—Cone 3—Pit Crater [24° 20′ 45.82″N; 39° 46′ 30.32″E]

Pit craters form due to sudden withdrawal of magma below a crater through flank eruptions leading to a fast collapse of the crater floor (Carter et al. 2007; Harris 2009; Németh and Cronin 2008). As a result the internal part of the crater wall will be mantled by draping lava and spatter. The outflow points are commonly marked as "boccas" in the outer edifice lower flank. Recognition of pit crater formation bears an important role to establish the eruption mechanism a volcano followed.

The sudden withdrawal of melt likely means that the crater was filled with active lava lakes and that was commonly acted as point source of low lava fountains (Okubo

Fig. 3.35 Reticulate and Pelee's hair and tear are abundant in the western edge of the lower section of the 1256 AD main cone flank [24° 20′ 51.92″N; 39° 46′ 17.31″E]

and Martel 1998; Rymer et al. 1998). The 1256 AD eruption site along the 2.3 km-long fissure shows numerous evidences of active lava lake formation then pit crater development. The repeated nature of pit crater formation attests the drainage and refill of magma to a crater. An example provides an ideal geosite to be defined as Cone 3 (Fig. 3.27). This crater is in a short walk from the main access point to the 1256 AD geotope and it provides a perfect view into a twin pit crater. In the inner wall of the crater lava spatters form ramparts and multiple layers of lava lake level markers suggest that the lava lake hosted in this crater changed its level more than once. This geosite has a high educational value to demonstrate that craters can form in a passive way, and explosive activity is not the only way to form a crater (Roche et al. 2001).

3.9.1.8 Geosite 8—Cone 4—Large Scoria Cone with Complex Crater [24° 20′ 52.39″N; 39° 46′ 26.66″E]

Cone 4 is one of the large volcanic cone of the 1256 AD volcanic geotope (Fig. 3.27). It is a complex scoria cone that exposes edifice sections dominated by lava spatter beds, while other sectors are more typical to pure scoria ash and lapilli accumulation. In the crater of the cone is complex and provides a good view to understand the gradual step-like growth of the cone commonly accompanied with magma withdrawal (Stovall et al. 2009) and small pit formation in the center of the cone (Fig. 3.36). The core of the cone is welded, and agglutinate layers tend to show slow plastic deformation features indicating that the heat of the high level lava in the crater and upper conduit welded the edifice significantly (Martin and Nemeth 2006; Sumner et al. 2005; Vespermann and Schmincke 2000). In the crater, large cracks suggest that the cone was slightly spread and tend to fall apart due to the hot base it was sitting on. The top of this geosites are ideal to have a look out on the 1256 AD eruptions and its extensive lava fields. The significance of this geosite is to demonstrate the potentially explosive nature of an eruption this type of volcanoes in the Harrat Rahat functioned in the past. The dominantly Hawaiian style eruptions in combination with the more typical Strombolian style eruptions suggest that this volcano was a complex volcano with complex eruption styles where the physical conditions in the upper conduit determined the style of eruption (Stovall et al. 2011).

3.9.1.9 Geosite 9—Inter-cone Ponded Pahoehoe Lava Field [24° 21′ 2.09″N; 39° 46′ 26.84″E]

Just NW from Cone 4, a very special area defines the geosites where ponded lava and its surface features can be studied (Fig. 3.27). This area is located on the eruptive fissure of the 1256 AD eruption site and it is a question if it is underlain by a feeder dyke or it is just a ponding feature in a depression between cones. Currently there is not enough data to decide this. The inflational features however are evident. The center is composed of shiny, smooth surfaced pahoehoe lava fields that are partially cracked and rotationally and/or vertically displaced. The cracks are commonly healed by more viscous lava (Fig. 3.37). The center part of the ponded zone is about 1–5 m below to its margin where dragged lava forming rampart like feature like a skin over an area indicating deflational force that collapsed the dm-thick crust of the lava pond (Harris et al. 2009; Head and Wilson 1989; Patrick and Orr 2012; Stovall et al. 2009). This region can be accessed by long walk through a rugged lava field and is only recommended to those can handled such conditions. The geological and educational value of this geosite is illustrating lava ponding, which is a common event with this type of volcanism (Anderson et al. 1999; Ball et al. 2008; Crown and Baloga 1999; Hoblitt et al. 2012; Parcheta et al. 2012; Self et al. 1998). Moreover lava ponding can host large volume of magma that can quickly be released that needs to be viewed as potential volcanic hazard (Patrick and Orr 2012).

Fig. 3.36 Crater zone of the Cone 4 demonstrate complex intra-crater processes in a growing edifice [24° 20′ 52.27″N; 39° 46′ 25.62″E]

Fig. 3.37 Ponded lava in between Cone 4 and 5 showing exceptional inflational and deflational features [24° 21′ 1.82″N; 39° 46′ 25.69″E]

3.9.1.10 Geosite 10—Lava Flow Cascade and Lava Flow Termination [24° 21′ 5.04″N; 39° 46′ 31.13″E]

A spectacular lava flow cascade is defined as a geosite just N-NE from the intra-cone ponded lava field explained above (Fig. 3.27). The geosite represents an about 500 m long confined lava flow that shows a perfect transition from the ponded pahoehoe lava flow textures to a typical slabby and rubble pahoehoe textures. The steep slope and the magmatic pressure from the ponded lava together are inferred to be responsible for the formation of a high speed cascading lava flow that broke apart the earlier formed lava crusts and carried them away from their original position (Duraiswami et al. 2014; Guest et al. 2012; Peterson and Tilling 1980; Rowland and Walker 1990). As a result a typical small-scale rubble pahoehoe lava unit formed. The geosite is unique because in a relatively short distance and easy walk the visitor can access the entire lava flow and see the flow transition very clearly. An extra speciality of this geosite is that the visitor can walk out to the end of individual lava flow units and examine the flow termination of this type of flow (Fig. 3.38).

Fig. 3.38 Typical lava flow terminus of a small, transitional lava flow [24° 21′ 4.44″N; 39° 46′ 32.28″E]

3.9.1.11 Geosite 11—Lava Flow Ponding and Draining Effect [24° 20′ 59.47″N; 39° 46′ 19.87″E]

In the other side of the ponded lava flow between Cone 4 and 5, another fine example can be seen to demonstrate the drainage effect of ponded lavas (Fig. 3.27). In this region the visitor can trace the lava tube that connected to the central ponded region, which has a collapsed and ripped off roof, probably caused by the mechanical erosion by the cascading lava flow already carried large rafts of lifted and rotated lava crusts. Along the open lava flow channels' margins, spatter levels mark previous lava level stages (Fig. 3.39).

3.9.1.12 Geosite 12—Cone 5—Bomb-Dominated Cone [24° 21′ 5.21″N; 39° 46′ 21.72″E]

Cone 5 is a cone nearly entirely composed of cannon ball-like bombs and lapilli slightly agglutinated in the flank of the cone (Fig. 3.27). The abundant lava cannonballs indicates that lava must have left the crater in larger packets and travelled through a significant length in the air, where surface tension and rotation of the bombs that formed the clasts now cover the entire flank of the cone (Fig. 3.40). The cannonball bombs and lapilli can be interpreted to represent the complex origins of the particles. The spherical shaped bombs and lapilli are commonly smoothed surfaced with textures indicate some rotational movement of the clast prior to solidification, resulting some degree of textural separation of the core and the rim of the particle closely resembling some sort of armoured lapilli or bomb texture. These can be interpreted as a sign that these pyroclasts originated when degassed magma ejected as separate fragments that tended to be pulled into lava spheres due to mechanical rounding upon rolling down fast in a semi-molten state on the growing edifice flank (Francis 1973).

However, cannonball lapilli and bomb with more uniform and smoothed rim and more dense core with entrapped vesicles or older (different textured) lava can be interpreted as recycled colder particle that were ejected subsequently by younger melt from a relatively stable lava lake constantly digested rolled back material and erupted them out through discrete explosions and/or fountain (Alvarado et al. 2011; Bednarz and Schmincke 1990).

The cone flank shows a spectacular view. This geosite can provide information to the visitor that active lava lakes must have existed in this crater, where recurrent bubble

Fig. 3.39 Unroofed lava tube with lava level markes in the preserved channel margin [24° 21′ 0.12″N; 39° 46′ 21.48″E]

coalescence exploded the degassed magma that then formed cannonball-like fragments. This is a more calm explosive processes in comparison to those where reticulate formed and therefore this geosite can provide a reference to two end-member style of explosivity a future eruption would likely cause in the Harrat Rahat. The crater of the cone is also well-exposed and provides further evidences for pit crater formation and crater floor subsidence by drainage of the lava lake.

3.9.1.13 Geosite 13—Cone 6–7—Main Cone [24° 21′ 17.45″N; 39° 46′ 18.82″E]

Cones 6 and 7 are considered to be the main cones of the 1256 AD eruption (Fig. 3.27). This double cone is the largest by volume, and it has a deep, elongated crater that can be traced over 400 m (Fig. 3.41). In the crater floor of the cones coarser crystalline lava represent the base of former lava lakes that solidified and crystalized slowly. Accessing the crater floor is challenging but not impossible and not necessary becausecircling the crater rim can provide insight to understanding the eruption mechanism of the main cone of the Al Madinah 1256 AD eruption. It is evident that that the elongated craters functioned as host of lava lakes. The lava lakes commonly changed their level, and they were likely drained in the NW outer flank of the cone edifice, where a small "bocca" looks as an initial point of the major long lava flows. The highest easterly crater rim and edifice composed of step-like crater rims that are separated by faulted zone of flank blocks suggest a gradual subsidence of the central part of the cone accompanied with some tendency to rebuild the cone through subsequent eruptive activity. This is a similar process as documented elsewhere and it seems that it is a common feature for a long-lived scoria cone that gradually emitted lava flows which is the case of this cone (Németh et al. 2011; Riggs and Duffield 2008). This geosite therefore can be used to demonstrate the causes and consequences of the edifice growth, rafting and lava flow initiation associated with long-lived scoria cones.

3.9.1.14 Geosite 14—Collapsing Cone Zone [24° 21′ 15.61″N; 39° 46′ 16.78″E]

In the western margin of the Cone 6–7, there are good examples to demonstrate the cone's instability during the

Fig. 3.40 Cannon ball-like lava bombs littered on the flank of Cone 5 [24° 21′ 5.03″N; 39° 46′ 19.64″E]

Fig. 3.41 Elongated crater zone of the main cone of Cone 6 and 7 [24° 21′ 21.70″N; 39° 46′ 23.93″E]

edifice growth (Fig. 3.27). In this crater rim, large (tens of metres wide) radially fractured zones can be traced, where agglutinated lava spatter beds tend to be displaced forming a mosaic-like pattern in map view (Fig. 3.42). This geosite is excellent to demonstrate the mechanical response of a still hot and semi ductile spatter bed to deform in a rigid fashion, and collapse towards the center of the growing pit crater. This geosite is unique and easy to access, and visitors could

Fig. 3.42 Radially jointed agglutinated spatter fragments gently dipping toward the pit craters provide good evidence to imagine the pit crater and crater wall collapse formation event [24° 21′ 14.78″N; 39° 46′ 16.80″E]

learn a lot how the edifice growth and destruction can play together in the shaping of the cone morphology.

3.9.1.15 Geosite 15—Lava Flow Field Slope Angle Changes, Flow Transitions [24° 22′ 30.41″N; 39° 46′ 5.46″E]

In proximal areas it has been demonstrated clearly in a relatively small-scale (hundreds of metres) that lava ponding and sudden break outs from the ponded zones can form transitional lava flow morphotypes. In more distal areas in the main artery of the 1256 AD 23 km-long lava flow there are very graphic examples to explore this phenomena in large scale. This geosite is one of the best examples to demonstrate that lava ponding can take place en-route along the main long lava flows, especially when morphology barriers or depressions are common. This geosite shows such ponded lava zones, that then quickly cascaded through an about 20 m drop in the topography, leading to form a channelized rubbly pahoehoe texture to develop on the fast moving lava flow. This geosite has an important role to demonstrate to the visitors that lava behaves very differently in comparison to water, and unexpected inflational events and ponding can occur frequently. When such ponded zones break out, fast moving transitional type lava flows tend to form. Thus this geosite provides a fine example of the paheohoe to aa lava flow transition as strongly controlled by the viscosity of the melt due to cooling and the slope angle on the flow move (Duraiswami et al. 2003; Kilburn 1981; Peterson and Tilling 1980; Rowland and Walker 1990).

3.9.1.16 Geosite 16—Lava Flow Squeeze Outs in Distal Areas [24° 26′ 23.07″N; 39° 46′ 17.45″E]

In the most distal areas of the 1256 AD lava flows unconfined lava terminus formed (Fig. 3.27). In these areas there are numerous small to medium-scale lava flow zones where squeezed out lava can be seen (Fig. 3.43). These squeeze out zones are important geosites as they provide some insight to the visitor that lava flows can be active long after the main flow body was emplaced (Rowland and Walker 1987; Sheth et al. 2011). This information is also important to volcanic hazard aspects of similar lava flows telling us that lava can stay in molten stage long after the emplacement of the flow.

3.9.1.17 Georoutes

The above listed geosites of the 1256 AD Al Madinah eruption are best to visit by following three suggested study paths (Fig. 4.44). There are three levels of study paths recommended to be developed. Two of them can be done by walking off from a general starting or access points, while the third one offer an introductory dirt road experience that follows a full circle around the 1256 AD cones and the medial part of the main lava flows of the eruption.

Northern Circuit Walking Path

The Northern Circuit walking path takes the visitor to the main cones—Cone 6 and 7—through a near complete circle (Fig. 3.45). Its access point is located in the northern edge of the main cones. The main goal of this walking path is to link

Fig. 3.43 Distal lava squeeze outs in the margin of the 1256 AD main lava flow [24° 26′ 33.07″N; 39° 46′ 24.28″E]

geosites demonstrate the eruption mechanism of the main, most obvious cone(s) of the 1256 AD eruption site. In addition this walking path provides unique vantage points to explore the proximal lava flow region that fed the 23 km long lava field.

Southern Circuit Walking Path

The Southern Circuit walking path is a slightly easier walking track in comparison to the Northern Circuit one (Fig. 3.46). The visitor by completing this study path will see the geosites demonstrate features associated with the lava inflation and deflation, formation of lava tubes and their collapses, and provide some information on the lava flow initiation through ponded lava regions. Because the access point for this walking track is nearby to a sealed road, this walking path can be separated into smaller segments that general physical condition visitors or school children can also complete.

Cone and Lava Field Car Route

The Cone and Lava field car route (Fig. 3.44) follows dirt roads that completely circle the 1256 AD eruption's 7 cones and their proximal to medial lava flow fields. This trip can be offered to visitors who are not interested in exploring the proximal areas by foot, and/or have limited time. This trip can provide multiple vantage points to the main cones of the 1256 AD eruption. En-route the circle provides short stop options to examine specific geosites, especially those associated with the lava flow surface textures. This study path also recommended as a test trip to those wish to explore the Harrat Rahat in its more remote geosites in the Precinct 2, 3 and 4.

3.9.2 Geotope of the 641 AD Historic Eruption Site

Al-Madinah City is located in a basin that is filled with thick alluvial deposits derived from the higher basement rocks standing as horsts blocking the western and northern side of the basin (Fig. 3.1). An historic eruption that took place in this basin in 641 AD is inferred to be sourced from four small volcanic cones (Fig. 3.47), aligned NNW—SSE (Fig. 3.48), located west of Al-Madinah City ($\sim$12 km from the Holy Mosque) (Camp and Roobol 1989; Moufti et al. 2013b; Murcia et al. 2015; Nemeth et al. 2013). This cones

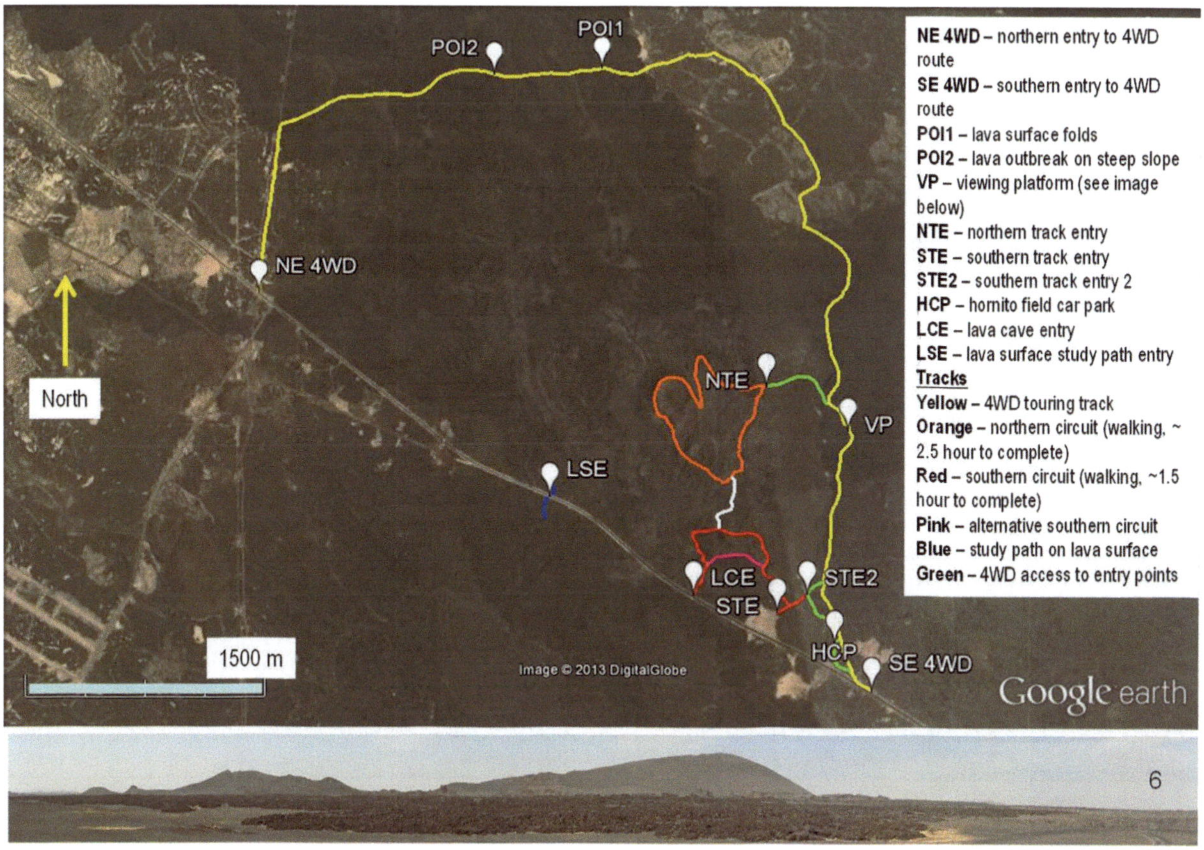

Fig. 3.44 Suggested study pathes across the 1256 AD eruption site of Al Madinah

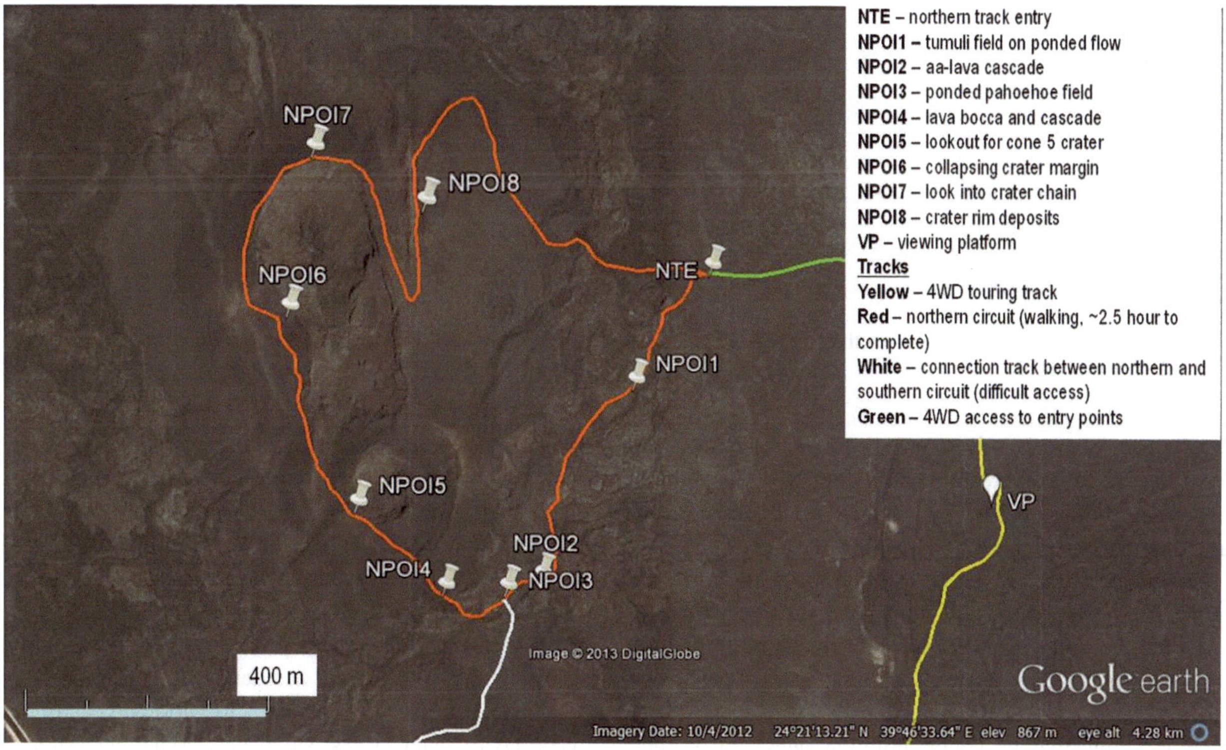

Fig. 3.45 Northern Circuit walking path

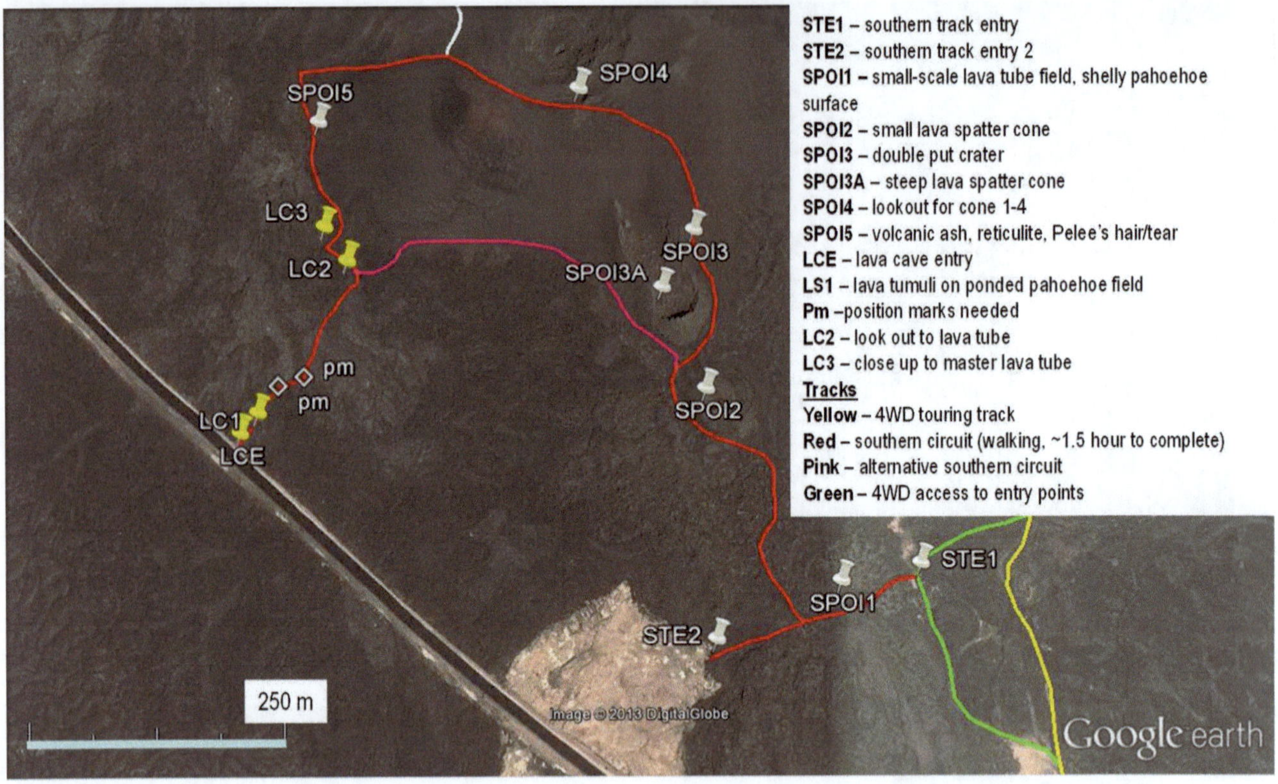

Fig. 3.46 Southern Circuit walking path

Fig. 3.47 Overview of the four cones of the 641 AD eruption site nearby Al Madinah city looking from the point of [24° 24′ 20.65″N; 39° 29′ 21.68″E] toward N

recently been named as the Al-Du'aythah volcanic cones (Murcia et al. 2015). Three out of the four cones have basal phreatomagmatic deposits indicating initial phreatomagmatic explosions (Fig. 3.49). These are the only centers of all the younger (<10,000 years) volcanoes in northern Harrat Rahat where evidence of phreatomagmatism is known.

The four cones have young volcanic morphology, such as steep cone flanks, near angles of repose slope, intact crater rims, and limited gully formation on its outer flank, all of which is suggestive of young eruption ages, when their geomorphology features are compared to other young cones elsewhere (Murcia et al. 2015). Each of the four cones is similar in size, with a base diameter of about 200–250 m and relative heights of about 30–50 m. The tallest, but simplest, volcano is the most southern and is composed of a conical shaped edifice with an enclosed single crater. The other three cones are somewhat more complex and exhibit multiple craters and complex volcanic stratigraphy, ranging from a basal tuff ring abundant in accidental lithic fragments commonly cored in lapilli and bombs to various types of scoria cones, lava spatter cones, small lava coulee and short lava flows. The upper sequences of the cones are inferred to be

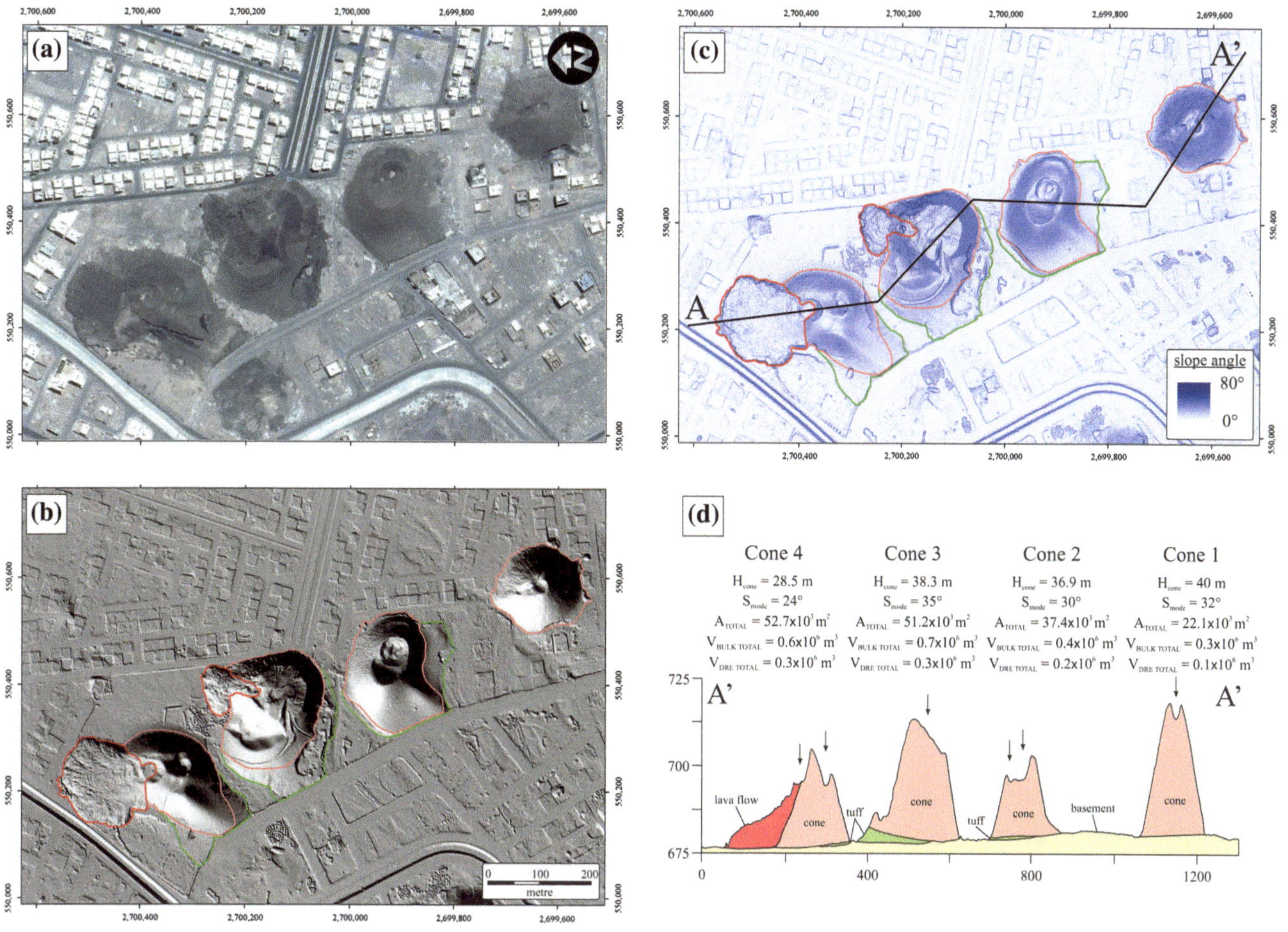

Fig. 3.48 Map view of the four cones inferred to be the source of the 641 AD historic eruption near Al Madinah city on GoogleEarth image (**a**), on LiDAR (**b** and **c**) and on a schematic cross section (**d**) with calculated eruptive volume values after Murcia et al. (2015)

produced by typical lava fountain and moderate Strombolian style explosive eruptions that also initiated small clastogenic lava flows reaching less than 300 m from their source.

The thickest tuff ring sequence has been recorded as a 5 m thick succession of lapilli tuff that is inferred to have been formed by an initial explosive eruption triggered by the interaction of rising basaltic magma and the shallow ground-water table, and the resulting pyroclastic rocks are defined as a basal phreatomagmatic succession. While phreatomagmatism is inferred to be the cause of the initial explosive, vent opening stage in many older (0.3–0.7 Ma old) volcanoes of the Harrat Rahat (Moufti and Németh 2013a), such records in association with small basaltic volcanoes are not known in the vicinity of Al-Madinah City, especially not in other young or historic eruption sites.

The basal tuff ring deposits also expose numerous cored bombs that are the spectacular results of the interaction between cold country rocks and low viscosity basaltic magma, capable of engulfing particles and being ejected as a cored bomb (Fig. 3.50). The fact that three out of the four cones have a basal tuff ring indicates that during this eruption, at least in the initial phase, magma interacted with shallow ground-water and triggered base surges that accumulated a relatively thin tephra unit (Rosseel et al. 2006). This also indicates that, if a future eruption were to occur in the area of the Al-Madinah basin, there is a chance that the initial stage of the eruption could be phreatomagmatic and therefore to promote and preserve this location as an intact volcanic geotope is important for volcanic hazard education (Moufti et al. 2013b).

The 641 AD volcanic geotope can host several individual geosites that are each can be used as a standalone geoeducational location to promote several aspects of mafic explosive and effusive volcanism. The proximity of the location to Al Madinah city, the site relatively small size and the complexity of volcanic features well preserved and exposed make this geotope a unique location future geoeducation programs could use, and as proposed could be the

Fig. 3.49 Basal phreatomagmatic tephra indicates phreatomagmatic explosive eruptions three out of the four cones formed during the 641 AD eruption [24° 24′ 48.47″N; 39° 29′ 44.88″E]

gateway to the great Al Madinah Volcanic Geopark (Moufti et al. 2013b). This geotope also contains the majority of the volcanic features the visitor can come across by visiting the greater Harrat Rahat region, and therefore it can be used as a jump-desk to develop any further geoeducational programs to a remote and fairly large region of Harrat Rahat.

3.9.2.1 Geosite—Cone 1—Intact Scoria Cone [24° 24′ 33.44″N; 39° 29′ 56.10″E]

Cone 1 is the most southerly cone of the four cones and it is also the simples (Fig. 3.51). It has a circular crater that is well-preserved. In the crater lava spatter banks exposed. In the flank of the cone is steep, typical to a scoria cone in spite that the majority of the edifice is composed of agglutinated lava spatter. To access the top of the cone needs some care as there is no path and the flank is steep. The geosite educational value is that the cone is still intact, its shape demonstrating a young volcanic landform and its pyroclastic rocks units indicates a relatively simple and short lived eruption mechanism (hours to days).

3.9.2.2 Geosite—Cone 2—Thin Veneer of Phreatomagmatic Base and Spatter-Covered Crater Interior (S3–4) [24° 24′ 42.18″N; 39° 29′ 51.17″E]

Cone 2 is a smaller cone than Cone 1 and its flank is not as perfect as Cone 1 (Fig. 3.52). The base of the cone in its southern edge are composed of a thin lapilli tuff succession indicating that the eruption of this cone must have started by an initial mild phreatomagmatic explosive phase that quickly evolved to be magmatic explosive with some intermittent lava fountaining events. The geosite educational value is to see clearly how an initial phreatomagmatic explosive eruption can turn to be more magmatic explosive and eventually build a scoria cone over the course of the eruption.

3.9.2.3 Geosite—Cone 3—Scoria Section (S1–1) [24° 24′ 49.63″N; 39° 29′ 48.40″E]

Cone 3 is a complex scoria cone (Fig. 3.53) with a phreatomagmatic lapilli tuff base that is covered by typical

Fig. 3.50 Cored bombs from the 641 AD eruption sites

Fig. 3.51 Cone 1 has an intact and simple cone [24° 24′ 33.44″N; 39° 29′ 56.10″E]

scoriaceous ash and lapilli upsection. The scoria section is important as it composed of vesicular typical scoria indicating that the eruptions were controlled by regular bubble outbursts in the upper conduit of the growing scoria cone. The black scoria contains numerous cannonball lapilli and bomb suggesting that some recycling must have been taken place in the crater where ejected bombs and lapilli fell back. The scoria section also shows some exhalation minerals that make the scoria deposits colourful in places. The geosites educational value is that it can help to the visitor to understand the nature of magma fragmentation and provides good evidence that ash can be produced in such small eruptions.

3.9.2.4 Geosite—Cone 3—Lava Dome (S1–2) [24° 24′ 51.50″N; 39° 29′ 49.08″E]

The top of the Cone 3 is covered by a lava dome and spine that fed a very short blocky lava flow. The lava flow is rather a lava dome that uplifted the internal part of the scoria cone

Fig. 3.52 Cone 2 has a broad crater mantled with lava spatter [24° 24′ 42.18″N; 39° 29′ 51.17″E]

Fig. 3.53 Cone 3 is a complex scoria cone with a phreatomagmatic lapilli tuff base that covered by scoria beds and lava spatter [24° 24′ 49.63″N; 39° 29′ 48.40″E]

and partially protruded through the edifice sliding fragments. This location provides a good example that volcano destabilisation and collapse can take place in such small volcanoes and they can pose a syn-eruptive hazard. This geosite also provides a good introduction to an intra-crater intrusive process that can be explored in large scale in the remote parts of the Harrat Rahat.

3.9.2.5 Geosite—Cone 3—Exposed Phreatomagmatic Base (S3–2) [24° 24′ 48.03″N; 39° 29′ 45.25″E]

The base of the Cone 3 is composed of about 4 m thick exposed lapilli tuff and tuff that is bedded, well-bedded to cross-bedded and contains abundant country rock fragments. Many of the country rock fragments are partially or fully covered by thin lava coat, indicating a low viscosity melt that entrapped them. The cored bombs are inferred to have been derived from the alluvial fan filling a basin nearby the bounding Precambrian horsts. The phreatomagmatic base is the thickest at the Cone 3 suggesting that the initial phase might have been short, but it has been excavated significant portion of country rocks that ended up in the accumulating basal pyroclastic succession.

This geosite bears with a very significant educational value to be able to show the differences of the pyroclastic succession formed due to explosive magma and water interaction. As phreatomagmatic successions are rare in the

Harrat Rahat, the presence of them in relationship with the small cones of the 641 AD eruption site can keep the public attention on the fact that in low-land and in more humid periods, phreatomagmatism can take place in an otherwise arid region. This fundamental volcanic hazard aspect cannot be underestimated.

3.9.2.6 Geosite—Cone 3—Exposed Transition Between Phreatomagmatic Base to Scoria Deposits (S3–3) [24° 24′ 47.41″N; 39° 29′ 47.58″E]

A transitional section in the southern flank of the Cone 3 can provide another unique geosite where the visitor can explore how continuous the transition between the phreatomagmatic and magmatic eruption driven pyroclastic succession. This indicates that the eruption were likely continuous (e.g. no break) and the eruption style must have changed during the evolution and growth of the cone. This geosite therefore can convey important messages such as (1) initial phreatomagmatic explosive eruptions can change to be more magmatic gas-driven if the water supply drops or vanishes in the course of the eruption and (2) if the magma eruption rate is large enough, the initial phreatomagmatic pyroclastic successions can be completely covered by a large scoria cone. This later message is important as other scoria cones, especially in the Al Madinah basin where ground water is available, might have had the similar thin phreatomagmatic base.

3.9.2.7 Geosite—Cone 4—Exposed Phreatomagmatic Base (S3–1) [24° 24′ 52.48″N; 39° 29′ 40.59″E]

Cone 4 is the most northerly cone of the four 641 AD cones (Fig. 3.48). The base of this cone similar to Cone 3 and this geosite demonstrate similar features but in different volcanoes as the previous geosite. The significance to of this geosite is that such thin phreatomagmatic veneers can form very irregular base of this cones that can be partially or entirely covered by subsequent eruptive products. Probably the Cone 3 and 4 are those cones where the magma eruption volume and rate were too small and the basal phreatomagmatic pyroclastic unit were not covered completely, and therefore we still can see that initial phreatomagmatic explosive eruption took place.

3.9.2.8 Geosite—Cone 4—Short Lava Flow Terminus (S2–1) [24° 25′ 1.91″N; 39° 29′ 40.92″E]

The Cone 4 is the source of a short lava flow that poured out toward the North (Fig. 3.54). This lava flow is short (<200 m) but still showing typical flow morphological features such as rubbly pahoehoe texture as it can be seen in many of the Harrat Rahat lavas. In this respect this geosite has geoeducational significance because it can be used as an introduction site to the lava flow emplacement processes common across the harrats of Saudi Arabia.

3.9.2.9 Geosite—Cone 4—Partially Collapsed Cone (S2–2) [24° 24′ 58.58″N; 39° 29′ 44.21″E]

The top of the Cone 4 is composed of a double crater. It seems that an initial crater has been truncated by the outpouring lava flow that partially rafter the flank of the cone away. As a result, the surface of the lava flow is littered by cone flank remnants and large pyroclasts such as fluidal bombs and blocks. This geosite therefore is important to demonstrate that the crater of a volcano is an active playground where gravitational collapse, rafting by outpouring of lava flows and explosive eruptions can act and shape it to their final form. This is again and important aspect to introduce in this geosites, because similar processes but in larger scale are very common across the Harrat Rahat, and therefore this geosite can be used as a starting point for such geoeducational programs.

3.9.2.10 Geosite—Cone 4—Crater of Cone 4 (S2–3) [24° 24′ 55.74″N; 39° 29′ 45.47″E]

The Cone 4 has a well-developed crater from where the visitor can have a nice view toward the short lava flow. From this point as a geosite the visitor can get an insight to the dynamic processes of how volcanic craters evolve.

3.9.2.11 Georoutes

Along the four cones of the 641 AD eruption site four study paths (Fig. 3.55) are recommended (Moufti et al. 2013b). These study paths are linked together and following the above described geosites. The basic concept suggested is that these four study paths combined with a visitor center could act as a gateway visitor center for the Harat Al Madinah Volcanic Geopark. Beside that the visitor center could act as a geoeducational center, the four study path can provide all the details in a short distance walk or drive that can give to the visitor a very good start to plan their more adventure-rich trips to the interior to the Harrat Rahat.

Walking Path 1

Walking path 1 is the shortest and easiest walking path exploring the Cone 4 (Fig. 3.55). The main goal of this study path to provide a first-hand easy experience to the visitor to see the conditions of a lava field, cone flank and the potential features they can come across in the harrats.

Fig. 3.54 Lava flow terminus of the Cone 4 [24° 25′ 1.91″N; 39° 29′ 40.92″E]

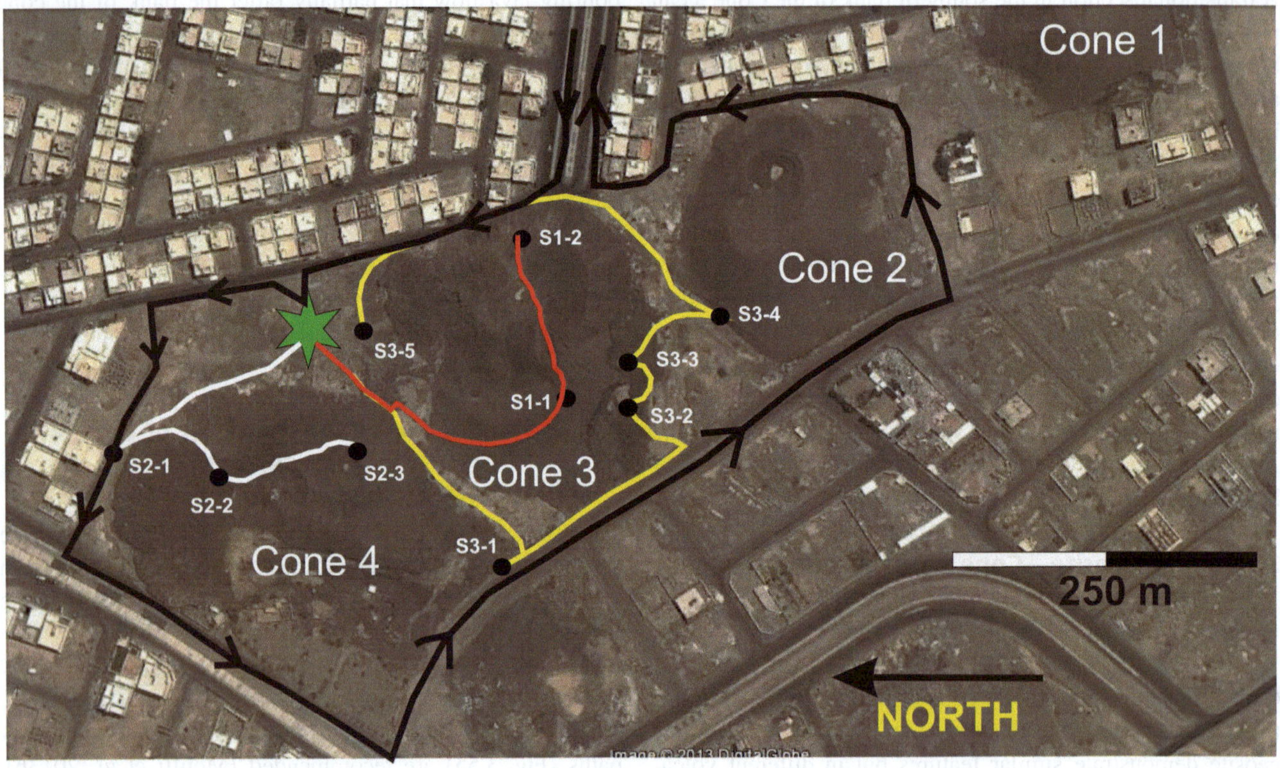

Fig. 3.55 Study paths of the 641 AD eruption sites. *Green* start marks the potential information center from where the walking pathes can be started

Walking Path 2

Walking path 2 requires a little bit better physical conditions from the visitor as it explores the Cone 3 (Fig. 3.55). It takes the visitor up to its summit and it also provides a good way to connect the basalt phreatomagmatic successions to the capping magmatic explosive and effusive units. From the top of the Cone 3 a perfect view can be enjoyed toward Al Madinah city. This walking path can emphasize the volcanic hazard aspect of the proposed geopark as the visitor will see clearly how much the city expanded since the 641 AD eruption, and today its outskirts are located in the harrat.

Walking Path 3

Walking path 3 takes the visitor around the Cone 3 (Fig. 3.55). Instead of walking up to the cone, the visitor can stay in level and just complete a circle to see the phreatomagmatic successions and the overview of the complex scoria cone that is capped by a lava spine.

Car Touring

For an easy overview, a driving tour can be arranged that takes the visitor around the four cones allowing stops at key geosites focussing on the basal phreatomagmatic succession

(Fig. 3.55). The car tour option also can provide opportunity for families to stop by and explore the volcanic features with good options for picnicking.

3.9.3 Scoria Cone with Ottoman Fortress Geotope [24° 20′ 17.47″N; 39° 35′ 13.14″E]

Just south of Al Madinah city around the ring motorway a complex scoria cone form an intact volcanic geotope (Fig. 3.56). This scoria cone seems to be a young scoria cone complex capped with historic site of an Ottoman fortress, and therefore it has a complex geoheritage value. The scoria cone was used as a perfect lookout point for centuries, and the fortress constructed on the top of the cone used the natural crater rim as a wall. The complex scoria cone also show evidences that it is a relatively young volcanic landform as based on its stratigraphy position in comparison to basal and capping lava flows nearby.

3.9.3.1 Geosite—Cone Base

The base of the complex cone demonstrates a fine example to examine the erosional processes of a scoria cone. The thick reworked scoria fan resulted from the gradual erosion of the cone itself, and its base now covered by modified grain flow-dominated volcaniclastic sediments. The way up to the cone follows the gradual transition from the reworked part of the cone to the more primary volcanic explosive eruption dominated scoria and bomb beds in the top of the cone. The geoeducational value of this geosite is based on its good exposure and graphic examples, that the scoria cone edifices' base are dominated by volcaniclastic deposits that were formed due to the mass movement on the flank of the cones.

Fig. 3.56 GoogleEarth satellite image of the complex scoria cone host an Ottoman fortress [24° 20′ 17.47″N; 39° 35′ 13.14″E]

3.9.3.2 Geosite—Cone's Double Crater and Fortress

The top of the geotope is an ideal place to see the volcanic craters of a scoria cone (Fig. 3.57). The crater rim is composed of agglutinated scoria especially around the lip of the crater. The agglutinated spatter-dominated beds of the crater rim were used to form the natural base of an Ottoman fortress upon the construction was based (Fig. 3.58). The collar-like crater lip also show a graphic example that the erosion of a scoria cone in such conditions are likely controlled by rock falls due to the undercutting of the softer scoria layers below the agglutinated more stable beds (Kereszturi and Németh 2012a, b; Martin and Németh 2006; Németh 2010; Németh et al. 2003, 2005).

3.9.4 Al Madinah Water Management Geotope

Around Al Madinah, water management in early history produced some spectacular and less known water management systems. Especially in the northern side of the greater Al Madinah region (Fig. 3.59), such remains are common, and they can serve good understanding on the water availability of the Al Madinah basin. This geotope plays and important role to demonstrate the ground and surface water is available, and probably was more abundant in the past historic time. This information can be connected well to demonstrate that phreatomagmatic explosive eruptions are likely eruption scenarios in areas where shallow aquifers are common. In this aspect this geotope fits very logically to the geoeducational concept of the Precinct 1 educational concept particularly to the demonstration of the formation of the 641 AD volcanic cones.

3.9.4.1 Geosite—Water Dam 1 [24° 26′ 34.15″N; 39° 55′ 8.38″E]

A large water dam form this geosite with a fantastic architecture constructed in a narrow gorge of old, Proterozoic rocks (Fig. 3.60). The water dam demonstrates that surface water can be captured and used for water management. In addition it is also an important site to appreciate the fact the in more humid climatic conditions even surface water can be abundant in the Al Madinah basin, that then can be diverted by lava flows and or trigger phreatomagmatic explosive eruptions at least in the initial stage of the eruptions as it has been demonstrated in the 641 AD cones just west of Al Madinah city.

Fig. 3.57 Double crater of the complex scoria cone indicates vent shifting during the eruption [24° 20′ 9.84″N; 24° 20′ 9.84″N]

Fig. 3.58 Agglutinated spatter collar functioned as a foundation of an Ottoman fortress demonstrates well the natural base of human developments on a scoria cone [24° 20′ 11.27″N; 39° 35′ 18.82″E]

3.9.4.2 Geosite—Water Dam 2 [24° 26′ 28.80″N; 39° 55′ 2.74″E]

A second major water dam show a smaller construct with lower but wider dam. The presence of this dam highlight the fact that in early time water management was fairly advanced in the region and surface water availability was great enough to invest to build such complex structures. This also indicates that especially in humid conditions, the potential to have phreatomagmatic explosive eruptions in the future if new vents would be opened in this region cannot be excluded.

Fig. 3.59 GoogleEarth maps of water dams just north of Al Madinah city

Fig. 3.60 Overview of one of the ancient water dam (1), part of the greater Al Madinah water management system [24° 26′ 34.15″N; 39° 55′ 8.38″E]

3.10 Precinct 2—Collapsing Cones, Lava Spatters and Lava Flows

This proposed precinct focuses on a single location that can be accessed only by 4WD tours. It is relatively far from Al Madinah city, but a well maintained 4WD track leads straight to the crater rim of a spectacular volcano (Fig. 3.61). On the way to the geosites of this precinct, the visitor will be able to see other lava spatter cones and extensive lava fields with tumuli, surface features and flat ephermal river and lake beds. The main geosites can be visited as a single full day trip or can be combined with a short morning visit to the Historic Eruption Precinct—1256 AD and 641 AD Historic Eruption Sites and then a visit to the Lava Lakes, Lava Fountains and Volcano Spreading Precinct—The Mosawdah Volcano.

Mosawdah volcano (24° 14′ 10″N; 39° 47′ 51″E; 1010 m asl) is a complex nested lava spatter cone with multiple crater rims, spreading fractures across the eruptive products surrounding the main crater (Fig. 3.62) and an at least 30 meters deep, perpendicular walled pit crater exposing a welded and rheomorphic lava spatter rim. The volcanic cone of Mosawdah volcano itself represents a single well-defined geotope where several individual geosites can be defined in accordance with the visible volcanological details. Each potential geosite is within walking distance of the other.

The volcano has been assigned to be ~0.6 Ma and 4500 BP in age (Camp and Roobol 1989), based on relative stratigraphy relationships with nearby volcanic landforms. At least three concentric nested crater rims can be identified around the main pit crater (Fig. 3.62). Each crater rim has a steep, near-perpendicular crater wall and relatively flat outward dipping outer rim. The entire volcano appears as a large (about 700 m wide) nested volcanic landform with multiple craters. Mosawdah volcano shows some similarities to the 1256 AD nested volcanic cones and logically can be connected to the geoeducational programs developed for the 1256 AD cones as part of the Precinct 1 of the proposed Al Madinah Volcanic Geopark. Mosawdah volcano however can provide a much more graphic example of an eruption that produced fast-moving, large-volume lava flows, high eruption rate driven lava fountaining and a complete rheomorphism of the accumulated pyroclasts along the active vents, similar to those that have been described during the Izu Oshima eruption in Japan in 1986 (Sumner 1998; Sumner et al. 2005). The Mosawdah volcano is open toward the northwest, from where a ~10 km long lava flow initiated toward the SW at about the same elevation from the outside flank of the cone as the base of the central pit crater.

The lava flows are relatively thin (few metres) tube fed pahoehoe flows and channelized a'a lavas that spread

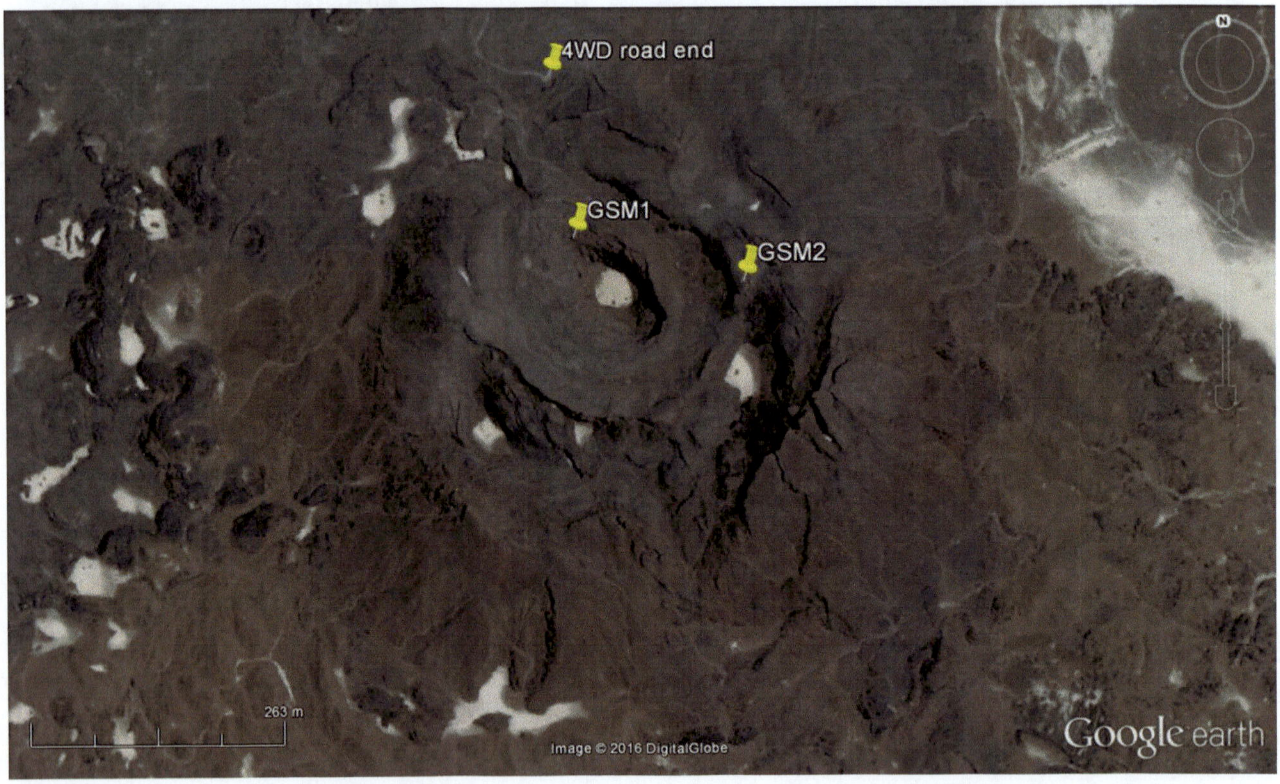

Fig. 3.61 Overview of the Mosawdah volcano and its surroundings on a GoogleEarth image [24° 14′ 10″N; 39° 47′ 51″E]. GSM1 and GSM2 refer to the location of the two geosites documented from Mosawdah volcano

Fig. 3.62 Overview of the crater of the Mosawdah volcano looking from the SE

broadly across the low lying areas around the cone. The lava fields are clearly visible from the top of this geotope and provide a spectacular view of a lava flow field that partially engulfs the central cone, leading to its gradual spreading and rafting. The main volcanic cone is about ~0.6 km in diameter at the base, with a maximum height of ~50 m, suggesting some sort of gradual spreading of the cone on top of a hot and fluid lava base.

Large lava spatter cones are common volcanic landforms associated with extensive low viscosity basaltic eruptions on intra-continental to ocean island settings (Kereszturi and Németh 2012a, b). There are numerous, well described examples of active lava spatter cone formation from Hawaii (Lefevre et al. 1991; Parfitt and Wilson 1995, 1999; Parfitt et al. 1995) or Iceland (Ilyinskaya et al. 2012). Lava spatter cone remnants are common volcanic landforms among many of the Miocene to Pleistocene European or western US intra-continental volcanic fields (Carracedo Sanchez et al. 2014; Valentine et al. 2000); however, they are commonly heavily vegetated and only sporadic outcrops of preserved rocks are visible. The Mosawdah volcano offers a perfectly exposed, non-vegetated, large volume example of the result of lava spatter eruptions. Mosawdah volcano also has a regional significance in terms of understanding the full spectrum of volcanic processes in the Harrat Al Madinah. The volcanism that created the Mosawdah volcano represents an end-member of the eruptive style spectrum, characterized by continuous and prolonged activity of relatively low lava fountains that provided fast accumulation of lava spatter around the active vent(s), promoting the formation of localized agglutinate and clastogenic lava flows.

The high heat source and the fast accumulation rate of lava spatter, in concert with a stable lava lake in the center of the volcano, created a ductile, partially molten base of the volcano, promoting gradual spreading and repeated lava lake drain-back and infill associated with extensive lava flow outbreaks. In this respect, the Mosawdah volcano is probably the best exposed and easiest to access site in the Harrat Rahat to provide an insight into the active lava fountain and lava lake driven eruption style that was common in the eruptive history of many of its volcanoes. Therefore, Mosawdah volcano and the proposed precinct around it bear significant geoeducational value to demonstrate the highly effusive, moderately explosive style of volcanism the region has experienced in the past and could experience in the future.

3.10.1 Mosawdah Volcano Geotope

3.10.1.1 Geosite—Pit Crater, Lava Outflow and Spatter Rampart [24° 14′ 13.80″N; 39° 47′ 48.37″E]

The Mosawdah volcano proximal crater rim forming successions are typical for fast accumulating lava spatters that locally form welded and clastogenic zones that squeezed re-melted material between individual lava spatter clasts (Sumner et al. 2005). The outline of lava spatters can be recognized; however, their recognition is becoming increasingly difficult toward the centre of the volcano (Sumner et al. 2005). This facies architecture indicates that the volcano erupted dominantly lava fountains, which must not have been very high (in the range of up to tens to few hundreds of meters) in order to be able to retain enough heat upon landing to allow the lava spatters to agglutinate and weld together locally, feeding clastogenic lava flows (Wolff and Sumner 2000). The near continuous section exposed in each of the preserved inner crater walls suggests no time break or interruption in lava fountaining (Fig. 3.63). The fast accumulation of lava spatters and the ongoing clastogenic lava flow formation must have generated an inferno in the proximal areas of the volcano, providing a soft, molten base to slowly slide and spread apart the cone itself, promoting repeated lava lake drawbacks, refill, volcanic sector dilation and partial cone rafting (Wolff and Sumner 2000). This geotope also shows a fantastic example to see how lava outpouring and pit crater formation can be connected. The central crater of the volcano is very deep and show evidences of plastic deformation of the accumulating spatter. The geoeducational value of this geosite is to highlight the fast rate of spatter accumulation needed to be able to form clastogenic lavas and therefore it provides a very important location to appreciate the significance of eruption rate in the formation and growth of scoria and spatter cones.

3.10.1.2 Geosite—Collapsing and Spreading Section of Cone [24° 14′ 12.73″N; 39° 47′ 53.48″E]

Due to the fast accumulation of hot pyroclasts and their agglutination and remelting to form clastogenic lava flows has a strong implication on the volcanic edifice stability (Németh et al. 2011; Sumner 1998; Sumner et al. 2005; Wantim et al. 2011). This geosite provides a graphic example to demonstrate this process and show how the volcanic edifice slowly spread apart along fractures (Fig. 3.64) over a hot still ductile deforming lava pond partially forming the lava lake in the growing crater. This site also poses another important aspect to form disintegrating scoria cones by potential bulging of shallow intrusive bodies due to the gravitational instability of large blocks

Fig. 3.63 Lava spatter dominated section of the crater wall area of the Mosawdah volcano [24° 14′ 11.31″N; 39° 47′ 51.55″E]

Fig. 3.64 Fractures likely formed due to the gradual spreading of the core of the growing volcanic edifice due to gradual outpouring of lava from the central lava lake and to the hot and plastically deforming base of the cone itself [24° 14′ 12.62″N; 39° 47′ 53.88″E]

of the core of the cone that can be uplifted by some upward moving melt packets in the base of the cone.

3.10.2 Al Anahi Volcano Geotope [24° 15′ 34.03″N; 39° 47′ 18.51″E]

Just north from the Mosawdah volcano, a remote scoria cone form a remarkable landform. The Al Anahi cone is a large scoria cone with complex crater and an extensive lava flow, all showing young morphological stages. The Al Anahi volcano can be defined as an intact geotope and can form an alternative site to visit for those wish to have adventure style geotourism.

3.10.2.1 Geosite—Al Anahi Cone [24° 15′ 34.03″N; 39° 47′ 18.51″E]

Al Anahi cone is a scoria cone that is among the largest cones in the Harrat Rahat (Fig. 3.65). The scoria cone is unique and it can be defined as a geosites on the basis that it has fairly high amount of fine ash and fine lapilli as an edifice building pyroclasts, and therefore suggests that its eruption was likely more explosive Strombolian style than pure Hawaiian lava fountain-dominated. The cone has an elongated crater that is easy to access by a short low grade ascend to the top. Around the crater rim the visitor can make a full roundtrip by walk, and see above the outflow point of a long lava flow partially rafter away small part of the cone.

3.10.2.2 Geosite—Al Anahi Flow Field [24° 14′ 51.53″N; 39° 45′ 3.01″E]

The lava flow of Al Anahi is defined as a geosite where the visitor can explore a thick and long lava flows with aa-type lava flow surface textures. The Al Anahi lava flow is one of the best examples in the Harrat Rahat to show typical aa lava surface textures. The lava flow is steep walled, contain abundant rugged lava spines, collapsed blocks and have milled cauliflower textures all indicative to more viscous

Fig. 3.65 Al Anahi scoria cone [24° 15′ 34.03″N; 39° 47′ 18.51″E]

lava to be emplaced. The flow itself acts as a major barrier to cross the harrat, and visits to this geosites are only recommended along its margins. In the marginal areas this lava flows show lots of evidences of late stage squeeze outs and collapse of the lava fronts exposing sheared conchoidal shape lava surfaces with pressure ridges. This geosite has a high educational value to demonstrate the lava morphology type varieties the lava viscosity can create, and also provide graphic example to demonstrate the volcanic hazard caused by a fundamentally aa-type lava emplacement.

3.10.3 Fissure Vent and Five Fingers Flow Field Geotope

One of the enigmatic locations in the Northern Harrat Rahat region is a chain of volcanoes along some visible surface fissures that look like a fissure source of some young lava flow field cited as "five fingers lava flow" (Fig. 3.66). The fissure oriented scoria and spatter cones are spectacular examples of a proximal source region of a major lava flows that initiated fine major lava flows filled valley networks in the NE side of the volcanic field. This region can be defined as a single major geotope on the basis of the unique link of its volcanic features demonstrate the evolution of a fissure aligned volcanic chain that emitted significant volume of fluidal magma through pahoehoe to aa lava flows.

3.10.3.1 Geosite—Fissure [24° 15′ 41.63″N; 39° 51′ 34.99″E]

An open fissure exposed in a lava flow field just west of a sealed road that can be defined as a key geosite to demonstrate the ground movement and fissure formation associated with volcanic eruptions (Fig. 3.67). The fissure is 2-10 m wide, deep, and exposes half section of whaleback type lava flows. The fissure can be traced over kilometres of length in satellite images and likely associated with some regional tectonic features along dyke movement took place. This geosite is very significant in geoeducational perspective as it can be shown to visualise the failed eruption took place in 2009 in the Harrat Lunayyir that caused seismic unrest and forced to evacuate thousands of people due to the fair of a volcanic eruption (Duncan and Al-Amri 2013; Jonsson et al. 2010; Koulakov et al. 2014; Mukhopadhyay et al. 2013). While the origin of this fissure is unknown its presence and its location in relationship with other fissure aligned vents

Fig. 3.66 "*Five fingers*" lava flow system on GoogleEarth image

Fig. 3.67 Open fissure cut through a thick lava flow field aligned in the same orientation as the scoria and spatter cones nearby [24° 15′ 41.63″N; 39° 51′ 34.99″E]

indicates that it had some link with dyke intrusions and eventual fissure eruptions that formed and shaped of the face of this part of the volcanic field.

3.10.3.2 Geosite—Southern Fissure Cones [24° 17′ 52.80″N; 39° 50′ 59.83″E]

The region commonly referred as the fissure vent zone (Murcia et al. 2014) can be divided into a southern and northern segment. This division is purely based on practical sense by the accessibility of the area. The southern segment is a very rugged region and to access this part needs great deal of orientation and potentially to visit this geosite would require organized tour. The fissure region when the visitor passing through, difficult to see due to the great variety of lava surface textures, and small spatter cones littered the terrain. The southern segment of the fissure aligned vent connected well with the exposed fissures explained in the previous geosite. Along a zone of about 7 km length what this geosite defines numerous individual scoria cones can be identified. Many of these scoria cones are rather chains of cones, and many of them have elongated craters. The southernmost part of this geosite is difficult to access, and only a look out can be taken from its southernmost locations toward a rafted and partially collapsed fissure network that is the inferred source of the southernmost arm of the five fingers lava flows. Further in the north the fissure aligned cones can be accessed on dirt roads, however, to navigate in the harrat in this rugged terrain needs lot of attention. Along a dirt road a spectacular spatter cone that looks like a giant hornito can be seen (Fig. 3.68), that acted as appoint source of lava fed from the southernmost arms of the five fingers flows. Along the fissure small lava flows forming a ribbon-like network of flow surfaces indicating the running nature of the low viscosity melt that must have just outpoured along this fissure (Fig. 3.69). Further to the north a complex and well-developed large lava spatter cone chain forming a fantastic landscape showing the dynamic nature of the entire fissure network feed lava over tens of km^2 areas.

3.10.3.3 Geosite—Northern Fissure Cones [24° 20′ 49.16″N; 39° 49′ 56.73″E]

The northern sector of the fissure aligned vent system in the Northern Harrat Rahat is the apparent source of the northern two arms of the five fingers lava flow (Murcia et al. 2014). In this location older cones form an obstacle for the outpouring of the lava however some remains of the original lava spatter cones can be recognized alongside with a great variety of collapsed pits and lava tube roofs. This geosite has a geoeducational value because it shows clearly that the rejuvenation of volcanic activity in a same place can form amalgamated volcanic landforms that overlap each other. The interaction of the vents emitted the lava flows with older cones aligned to the same direction than the fissure suggests that the region was common area where volcanic eruptions took place. This site therefore can contribute significantly to our understanding of the volcanic hazard aspect of a long-lived volcanic field.

3.10.3.4 Geosite—Five Fingers Lava Flow Field Proximal Section—South [24° 20′ 49.16″N; 39° 49′ 56.73″E]

The proximal section of the lava flow fields of the five finger lava flow is an important geosite. While on the satellite image and on the field the region is an area where only thin lava flow preserved, it can provide a very important eruption mechanism aspects to the visitors. Interestingly in these regions only the marginal chilled crust of a former inflated flow system is preserved, and scattered lava crust rubbles, that form a field with thin obscured lava blocks (Fig. 3.70). The preserved lava crusts and marginal fragments with some squeeze out flows suggest that the majority of the flow must

Fig. 3.68 Giant hornito-like spatter cone feeding fluidal (low viscosity) lava flows [24° 20′ 0.08″N; 39° 51′ 2.59″E]

Fig. 3.69 Low viscosity lava formed amazing lava network on the surface of a fissure-fed spatter cone. Individual lava tube is about a meter wide

Fig. 3.70 Proximal region of the southern segments of the five finger lava flow

have been removed gravitationally. Such situation can be envisioned that a ponded lava flow formed nearby the fissure that inflated, and then when it has reached a threshold volume value it has been collapsed and emptied its fluid interior fast, that has carried away the broken crustal regions. This geosite provides a very graphic image for this potentially dangerous process we need to consider in eruption scenario developments for future lava flow eruptions in the region.

3.10.3.5 Geosite—Five Fingers Lava Flow Field Proximal Section—North [24° 21′ 12.61″N; 39° 49′ 14.15″E]

In the northern sector the proximal lava flow regions of the five finger flows show large scale hummocky surface morphology (Fig. 3.71). These features can be interpreted as an inflated flow field that has been partially preserved, e.g. they have not been removed by gravitational collapse and drag of the fast emptying of the ponded lava. In this respect this geosite is an important one to compare the lava pond emptying processes that can be seen in detail in these sites.

3.10.3.6 Geosite—Five Fingers Lava Flow Field Median Section [24° 24′ 56.11″N; 39° 50′ 56.18″E]

The medial sections of the five fingers lava flows show typical transitional lava flow morphotypes (Fig. 3.72). This geosites all along the middle sections of any of the five fingers five arms can be visited with an aim to demonstrate the nature of the pahoehoe to aa flow transition in the form of ripping off solidified lava crusts that then homogenized and mixed with the moving lava flow main body. This geosite can provide a very insight to the visitor about the flow transition and the gradual ponding and breakout as an important movement mechanism of transitional type lava flow movement.

3.10.3.7 Geosite—Five Fingers Lava Flow Terminus [24° 23′ 42.79″N; 39° 59′ 18.36″E]

The terminus of the five fingers lava flow five arms provide a good example to envision how a transitional, more aa type lava stops in the end (Fig. 3.73). This geosite also very graphic example to imagine the momentum of such a long (over 10 km) lava flow can have and how it come to a rest by forming an unconfined fan in its end that hit the pre-existing topography.

3.10.3.8 Geosite—Five Fingers Lava Flow Field's Rafted Cone Material and Northern Flow Terminus [24° 26′ 42.78″N; 39° 50′ 23.48″E]

One of the most intriguing textural feature of most of the lava flows in the Harrat Rahat is the potential of this type of lava flows to carry bulldozed flow material as well as pre-existing cone material as rafted debris (Fig. 3.74). This geosite has a significant message for volcanic hazard of large lava flows inundate inhabited human built environment, as they can move building fragments far, and apparently they cannot be stopped easily by obstacles (Fig. 3.75).

3.10.4 Zargat Abu Zaid Geotope [24° 16′ 32.83″N; 39° 50′ 25.16″E]

Zargat Abu Zaid is a large scoria cone just west of the fissures fed five fingers lava fields. The cone is very likely an older cone as its flank has extensive gully network in spite of its steep cone flank. The crater rim is well-preserved as a collar-like feature as a result of the preservation of softer scoria beds capped by lava spatter banks. This scoria cone is

Fig. 3.71 Whaleback style lava morphology of a large inflated and preserved flow part of the source region of the northern sectors of the five finger lava flows [24° 21′ 12.61″N; 39° 49′ 14.15″E]

Fig. 3.72 Pahoehoe to aa transitional lava flow morphotypes in the medial sections of the five fingers lava flow [24° 24′ 56.11″N; 39° 50′ 56.18″E]

Fig. 3.73 Lava terminus of a nearly 15 km long lava flow [24° 23′ 42.79″N; 39° 59′ 18.36″E]

a perfect example to demonstrate the syn-eruptive eruption style influences on the volcanic facies architecture that then controls the post-eruptive erosion style of the cone (Kereszturi and Németh 2012a, b). The scoria cone also has an extensive lava field that partially rafter away its northern sector that have been partially rebuilt by the ongoing eruption. This scoria cone with its lava field can demonstrate the cone rafting processes that took place in a time when the cone was still erupting and producing ash and lapilli that then partially covered the moving lava flow. The cone with its lava field together is a geotope due to the intact geological features they can demonstrate.

Fig. 3.74 Rafted nearly completely homogenized cone material (lighter textured fragments among darker coherent lava clasts) incorporated in the aa lava in the distal facies of the five finger lava flow arms [24° 26′ 42.78″N; 39° 50′ 23.48″E]

3.10.4.1 Geosite—Crater Infill [24° 16′ 30.02″N; 39° 50′ 17.88″E]

The crater of Zargat Abu Zed scoria cone is partially filled with reworked volcaniclastic material as a sheet wash deposit accumulated in the crater of the cone (Fig. 3.76). The crater rim composed of agglutinated and partially welded lava spatter that acted as an erosion resistant cover protecting the cone from significant erosional lowering. Along the crater rim outer margin the erosion is dominated by rock falls due to undercutting of the solid and erosion resistant agglutinate. In the inner part of the crater steep dipping agglutinate beds forming the upper section of the crater that gradually covered by volcaniclastic sediments in the lower section of the crater. The northern edge of the crater rim is truncated and repeatedly rebuilt during rafting event of lava flows exited the lower northern flank of the cone. This geosite has a high geoheritage and geoeducation value because it is an easy to access location where the visitor can learn great variety of volcanic processes. In addition the geosite also can provide a good reference to show to the public that the final shape of a volcano can be the result of a combination of primary, edifice building and secondary, syn- and post-eruption related erosional processes.

3.10.4.2 Geosite—Rafted and Ash-Covered Flow Field

The northern sector of the Zargat Abu Zed scoria cone has been partially destroyed and partially rebuilt by subsequent eruptions. The northern sector of the cone composed of hummocky surface that is primarily a lava flow (Fig. 3.77) field initiated from the lower sector of the cone flank through a lateral "bocca". The lava flow is nearly completely covered by thick ash and lapilli nearly in its entire length making difficult to see the original surface and texture of the lava flow. The lava flow is initiated from the lower part of the cone and its movement is likely rafted part of the cone away. Larger cone sections on top of the lava flow is supporting this idea however the thick ash cover on the lava flow hinder further, more detailed interpretation. The fact that the lava flow field also covered by ash and lapilli suggest that the time the lava flow was still moving the scoria cone still had an explosive ash- and lapilli-producing eruption. This geosite has a high geoeducational value because it helps to link the cone (edifice)-building eruption styles with those forming the lava flows. Visiting this geosite and the entire Zargat Abu Zed geotope can help the visitor to recognize the link between explosive and effusive processes associated with the

Fig. 3.75 Zargat Abu Zaid scoria cone rafted cone and ash covered lava field. Photo was taken from the point of [24° 16′ 32.70″N; 39° 50′ 19.38″E]

Fig. 3.76 Complex volcaniclastic sedimentary infill in the Zargat Abu Zaid scoria cone's crater looking from the point of [24° 16′ 30.02″N; 39° 50′ 17.88″E]

harrats' volcanism. A geotope like this can serve well to develop a realistic volcanic image in the mind of the visitor in what the various volcanic processes are linked and not presented only as individual singular events and processes.

3.10.5 As-Sahab Geotope [24° 21′ 16.42″N; 39° 48′ 48.00″E]

As-Sahab is large and complex slightly older scoria cone that is in the region where the five fingers lava flow fields erupted along the main fissure system (Fig. 3.78). The cone complex has no assigned lava flow fields or at least it is too difficult to link the many old lava flows located nearby this cone complex to be able to say confidently which one belongs to the cone. The cone together however can be looked at as a complex geotope due to its special location. The cone acted as a barrier and obstacle for the five fingers fissure eruption and its lava flows were diverted by the cone itself. This location can show a perfect site to the visitors to appreciate the effect of a pre-existing topography in the potential path of subsequent lava flows. This geotope also poses a

Fig. 3.77 Rafter lava flow field of Zargat Abu Zaid with ash cover looking toward Al Madinah City (north) from the crater rim of the cone

Fig. 3.78 As-Sahab scoria cone complex on a GoogleEarth image. Top left corner of the map view is [24° 21′ 47.07″N; 39° 48′ 6.99″E]

fundamental question the visitors could see clearly, that older cones exists in the area where young fissure-fed lava flows erupted, moreover they are aligned in the same way as the young fissures and flows, and seemingly erupted in the same style. This has a significant message and high geoeducational value as this geotope can show very graphically that the NNW—SSE dorsal zone of the Harrat Rahat was/is one of the main eruption site of the volcanic field, and it is very likely structurally controlled as modern fissures and associated cone chains indicate it.

3.10.5.1 Geosite—Cone Complex

The As-Sahab scoria cone is a large cone in its advanced erosion stage (Fig. 3.78). The cone southern side is open and the cone has been eroded and lowered significantly. The geosite can be accessed by foot through the southern open crater. In the interior of the crater the inner crater wall is filled with lava spatters forming hanging lava drapes on the agglutinated steeply crater-ward dipping beds. Large clastogenic lava zones can be seen from below, however, access to those sites and in general to the crater rim's higher part is not advisable due to high potential for rock falls. The crater internal part is partially filled with volcaniclastic deposits mixed with aeolian dust. In spite of the infilling, the crater is still forming a shallow depression suggesting that its original depth and geometry must have been dramatic, and likely hosted a lava lake that have been drained laterally by "boccas" feeding the older lava flow fields located in this area. The significance of this geosite is that it can show a link between young and older volcanic landforms and demonstrate clearly that similar volcanic processes must have taken place over thousands and thousands of years. This geosite also provides a good example to compare the older and younger volcanic landforms.

3.10.5.2 Geosite—Interaction Between Fissure Flow and Cone

Jabal Al Malsa cone northern and eastern sector is partially covered by the five finger fissure-fed lava field (Fig. 3.78). Previously the cone was under debate as a potential main source of the fissure fed young Quaternary lava flows nearby. Recent fieldworks however suggest that Jabal Al Malsa is an older cone due to its morphological appearance. Along the contact between the young fissure-fed lava flows and the cone it is perfectly visible how an obstacle can divert and interact with a pahoehoe to aa transitional lava flow field. Along this contact zone the visitor can see complex lava surface textures while in the distant regions the proximal lava flow fields show a stunning volcanic landscape with open channels, pressure ridges, inflation features, tumuli and large rotate blocks of ripped up lava crusts. The geosite geoeducational value is high as it is also provide a very good vantage point to demonstrate that the older scoria cones in the region are also aligned to the same fissure orientation along the younger extensive lava flows were emitted.

3.10.6 Halat Khamisah Scoria Cone and Lava Flow Field Geotope [23° 55′25.18″N; 39° 54′38.53″E]

The Halat Khamisah scoria cone is defined to be a young scoria cone located near a major trans-Arabian gas pipeline (Fig. 3.79). The cone and its lava flows are remote and only dirt road access is available that might change in the future when a service road along a major gas pipeline is constructed. It is located in the central part of Harrat Rahat, and it can be selected as a type locality to demonstrate that Harrat Rahat had some other young volcanic region where scoria cones and lava flows erupted probably in the same time as the five fingers lava flows near Al Madinah. The cone and its single lava flow are together defined as a geotope on the basis of its volcanologically common features that can tell a unique story to the visitor. The geotope has a great educational value as it can demonstrate that young eruptions are not restricted to specific northern regions of the Harrat Rahat but they are known elsewhere. This geotope also can show that extensive ash-producing eruptions are also not restricted to the 1256 AD eruption but similar eruptions took place elsewhere. This geotope also can provide insight to understand the complexity of the evolution of Harrat Rahat and appreciate its volcanism's time and space dispersion.

3.10.6.1 Geosite—Ash Plain [23° 55′39.00″N; 39° 54′48.01″E]

The Halat Khamisah scoria cone forms a fairly large scoria cone that has steep cone flank indicating its young age. The edifice is covered by dark (black) scoria ash and lapilli that is littered by well-developed spindle bombs especially in the proximity of the cone. Where the cone slope angle reduced, black ash plain can be traced over few kilometres away from the cone foothill. This ash plain is exposed in cross-sectional view in few wadis about a km away from the cone, providing unique window to the internal texture, thickness and vertical variations of a scoriaceous ash and lapilli succession. The ash plain as a geosite beside its aesthetic value provides an important insight to understand the explosive eruptions associated with the cone building. The extensive nature of the ash plain suggests that at least in few stages, the eruption of the Pipeline cone might have been more violent than a normal Strombolian style explosive eruption, and it has been able to produce large volume of finer grained pyroclasts that were dispersed across the landscape as documented elsewhere (Kawabata et al. 2015; Németh et al. 2011; Pioli et al. 2008; Rowland et al. 2009). This geosite has a great geoeducation value as it demonstrates that in Harrat Rahat violent Strombolian style eruptions took place and we have every reason to assume that such eruptions might take place in the future.

3.10.6.2 Geosite—Cone Crater [23° 55′11.93″N; 39° 54′28.45″E]

The Pipeline cone crater can be accessed relatively easily from the cone southern foothill. The walking path to the crater is a gentle ascending path going through various lithofacies of the edifice building pyroclastic units. There are large spindle bomb-rich parts and fine grained more scoriaceous lapilli- and ash-dominated successions suggesting that the eruption style of this cone through its growth has changed many times. The crater of the cone is partially open toward the west and it is occupied by an outflow of a major blocky—aa lava. In the breaching point of the lava flow the cone flank is destroyed and displaced by rafting, however rafter materials cannot be traced long from the breach point suggesting that the crater rim might have been originally low in the west. The proximal zone of the lava filling the crater is typical aa lava with spines, large lava columns and lifted and rotated angular, crystalline lava blocks (Robert et al. 2014; Wantim et al. 2011). This geosite has a reasonable geoeducational value as this is among those rare sites where typical aa and block lava is exposed from their proximal section. Pahoehoe lava surface types are rare in this lava flow field.

3.10.6.3 Geosite—Block Lava Flow Field [23° 54′ 58.21″N; 39° 53′ 3.59″E]

This geosite refers to a lava flow cross sectional point about 3 km from the crater down-flow. Going up to the crater and follow the lava flow down is a very challenging track and it is not advised as the danger of injury is high due to the numerous steep and deep fissures between lava blocks. In

Fig. 3.79 Halat Khamisah scoria cone and lava flow is a young volcanic region in the southern central part of Harrat Rahat clearly visible by its young lava flows shown up with different textures on a GoogleEarth image. Top left corner of the map view is [23° 56′ 43.04″N; 39° 46′ 45.92″E]

this geosite however it is clearly visible how a typical aa-lava looks like. The lava is composed large blocks, and in places it closely resembles block lavas rather than aa. The lava flow interior has numerous large rounded cauliflower type lava balls, and vertically oriented spines as typical features of aa lavas. This geosite geoeducational value s to demonstrate that a seemingly block-dominated lava that is generated from a higher viscosity melt (more crystalline for instance) can produce lava flows that can travel over 10 km from their source even in a relatively flat surface.

3.11 Precinct 3—From Silicic Lava Domes to Explosion Craters

Explosive volcanic processes are typically the most hazardous aspect of volcanism to human life and associated built environments of the modern society. Explosive volcanism registers on a broad scale from weak to highly explosive eruption styles.

The Precinct 3 at the proposed Harrat Al Madinah Volcanic Geopark named as "From Silicic Lava Domes to Explosion Craters precinct" offers a very dramatic insight for the visitors into the eruptive products that result from various types of explosive volcanism from different magma compositions (Fig. 3.16).

Explosive volcanism associated with monogenetic intra-continental volcanic fields is typically caused by magmatic gas expansion of volatile-rich magmas, commonly more silicic in composition, and/or by magma-water interaction, causing phreatomagmatic explosions that form base-surges and other pyroclastic density currents and construct maars and tuff rings (Valentine and Gregg 2008). The central part of Harrat Al Madinah contains the best closely spaced examples to see the results of explosive volcanism in the form of extensive pyroclastic density current deposits and broad and deep explosion craters (Camp and Roobol 1989). These violent types of volcanism are inferred to be several hundreds of thousands of years old (Camp and Roobol 1989), seemingly forming a concentration of specific volcano types in a remote but accessible region of the proposed volcanic geopark. As it is the furthest precinct from Al Madinah city, as well as it contains some of the oldest volcanism in the Harrah Al Madinah, it is logical to offer this precinct as the last for the visitor. The logistical difficulties in visiting this precinct also make this location suitable for the more adventurous tourist with higher levels of fitness (Moufti and Németh 2013a). However, visitors to the precinct will be rewarded with probably the most dramatic, and one of the most unique, volcanic landscapes anywhere on Earth.

As is typical of arid areas with limited ground and surface water availability, the Al Madinah Volcanic Field has dominantly produced scoria cones, spatter cones and large lava flows, all derived from "dry" magmatic eruptions as presented in the previous two precincts (Moufti and Németh 2013a). However climate changes over the lifespan of a volcanic field (millions of years) can dramatically change the hydrology and hydrogeology of the region, and an otherwise "dry" eruption style-dominated field can quickly can be switched to a "wet" eruption dominated system, even without requiring dramatic magmatic composition changes (Kereszturi et al. 2011). As a result such volcanic fields can produce phreatomagmatic volcanoes such as maars and tuff rings.

Beside the dramatic volcanic landscapes of the explosion craters, the Precinct 3 includes truly unique volcanic landforms that record silicic lava dome formation and associated features. It is an interesting aspect of the Harrat Al Madinah's volcanism, which has global significance, that silicic (mostly trachytic) lava domes have been produced in the same region where basaltic scoria and lava spatter cone forming eruptions were dominant (Moufti and Németh 2013a). There has been a diverse range of silicic volcanism through non-explosive lava dome eruptions to block-and-ash flow generating violent eruptions (Moufti and Németh 2013a). Seeing these features coexist with features from basaltic and trachytic monogenetic volcanism is one of the most geologically intriguing aspects of the proposed geopark, in terms of understanding the origin of the evolved volcanism in dispersed, intra-continental systems commonly referred to as monogenetic fields.

In the proposed precinct, three well distinguished and easy to access trachytic lava domes as individual volcanic geotopes are included in the selected list of high value locations. However, from these lava domes such as local high points of the region, further distant lava domes can be seen toward the east and northeast as completing the picture for the visitor to appreciate that the region is very rich in silicic lava domes (Fig. 3.80). The geotopes and their geosites of the Precinct 3 provides very graphic insights to the physical processes associated with lava dome formation to be compared, with cryptodomes through to explosion craters, and complex volcanic structures showing evidence for multiphase, and commonly multi-chemical (basaltic to trachytic), eruptions.

The variety of trachytic lava domes exposed in is precinct are spectacular and they offer a good geoeducational program to build on them to link this geotopes and geosites to other well-known, recently erupted silicic lava domes, including Unzen in Japan (Kaneko et al. 2002; Yamashina and Shimizu 1999) and Mount St Helens (USA) (Anderson et al. 1995). The similarity in size, volume and eruptive products of the trachytic lava domes of the Harrat Al Madinah to those lava domes generally associated with subduction-related strato- and/or composite- volcanoes, makes the HAMVG truly unique and globally significant; and can make the proposed Harrat Al Madinah Volcanic Geopark as a potential "*Makkah of Volcanologists*" (Moufti and Németh 2013a). The scientific value of this precinct is self-defined. The perfect exposures, the lack of vegetable cover, the great visibility and the numerous longitudinal sections along gullies offer great research potential in these locations to scale the physical parameters of pyroclastic flow forming eruptions. In addition, the availability of pyroclastic deposit-engulfed scoria cones and other morphological obstacles can help to calibrate the energy budget of pyroclastic flows and therefore the precinct could serve as an important study location for such scaling volcanological work. Considering the "monogenetic volcanism" on display, this precinct offers a dramatically new view that will enable visitors to appreciate the complexity of such volcanism and see the link between focused (strato- and composite volcano-producing) and dispersed (purely monogenetic volcano-producing) magmatic plumbing system-associated volcanism, which will potentially make this volcanic geopark globally very significant (Moufti and Németh 2013a).

3.11.1 Matan Lava Dome Geotope [24° 13′ 31.71″N; 39° 50′ 23.56″E]

In the NW entry to the precinct, a dirt road follows a dry valley that connects to a broad alluvial fan which is bordered by the large Matan lava dome (Fig. 3.81). It is a complex volcano with a basal diameter of about 1.8 km. The dome is clearly a composite lava dome (Fig. 3.81) recording multiple styles of silicic magma emplacement that range from rigid to more plastic emplacement. Lava dome rock facies are diverse as a reflection of variously degassed and viscous trachyte being emplaced in a relatively confined area. Lithological domains can be seen in outcrops and they are commonly reflected by colour differences in the exposed rocks. The lava dome itself is dominated by coherent lava dome cores with some rock-fall and lava dome carapace breccias, some short run-out distance (<500 m) and relatively thin pyroclastic flow deposits can be traced around the main body of the lava dome complex all typical for a lava dome anywhere (Brenna et al. 2012; Costa et al. 2012; de Vries et al. 2014; Fink and Griffiths 1998; Miallier et al. 2010; Riggs et al. 1997). On the SW side of the lava dome, a volcanic crater with a flat crater floor suggests some initial explosive volcanic eruptions prior to the emplacement of the main Matan lava dome complex. This explosion crater is the source of at least two pyroclastic flow units traceable over a kilometre from the preserved

Fig. 3.80 Distant silicic lava domes dominate the horizon in the southern margin of the proposed Harrat Al Madinah Volcanic Geopark demonstrating that the visitor has entered to a region the intracontinental volcanism produced a great variety of volcanic landforms [24° 13′ 19.65″N; 39° 50′ 59.90″E]

crater rim. This location not only provides a graphic example of trachytic lava dome formation, but it also links the trachytic explosion craters with trachytic lava dome forming events, giving an example of the interplay between dramatically different eruptions styles.

3.11.1.1 Geosite—Matan Lava Dome Side [24° 13′ 12.97″N; 39° 50′ 2.40″E]

This geosite is the base of the lava dome complex in the southern side of the volcano. To access higher portions of the lava dome require some rock climbing skills and still it would be a difficult and demanding climb. In the base of the lava dome however a great variety of large blocks are ready to be observed and they can provide enough information to understand the lava dome-forming processes. Large blocks of flow banded trachytic rocks are the most common rock type among the debris apron. These type of rocks are typically representative for the lava dome growth and they can form zones across the original cliff faces representing specific parts of the lava dome complex. From the distance the shape of the lava dome is noteworthy especially its steep upper segment (Fig. 3.82). There are no obvious pyroclastic rock units exposed around the lava dome complex suggesting that lava dome collapse and associated explosive eruptions were plazed just a minor role in the evolution of this volcano. The geoeducational value of this geosite is that the visitor can get an immediate insight of the highly viscous nature of the magma formed this volcano. This view can be contrasted by nearby basaltic spatter cones and lava flows which are much smaller and more flat volcanic landforms. In addition, this location is in easy access from a sealed road.

3.11.1.2 Geosite—Matan Lava Dome Side Crater [24° 13′ 11.75″N; 39° 50′ 5.15″E]

In the southwestern side of the Matan lava dome complex a shallow, broad crater is visible that is surrounded by a low pyroclastic rim. This location can be accessed by walking track from the previous geosite through a gentle ascending path that take the visitor to the southern crater rim's top. Along this walk, the visitor can examine the debris apron surrounding the main lava dome, a typical volcaniclastic debris apron that is composed of short run-out distance block-and-ash flow deposits mixed with fluvial and

Fig. 3.81 The Matan lava dome geotope from the air showing a superb complex lava dome easy to access

Fig. 3.82 Matan lava dome complex from the geosite located in its southern foothill. Photo was taken from [24° 13′ 12.97″N; 39° 50′ 2.40″E] toward the NE

aeolian deposits and rock falls from the main dome region. The crater itself hosts voclaniclastic debris from the western segment of the Matan lava dome suggesting that it was formed earlier than the main lava domes. The crater rim is composed of welded pyroclastic rocks interbedded with some unconsolidated trachyte fragment-rich pyroclastic succession interpreted to be small-volume block-and-ash flow deposits. There are no clear indications to support magma and water explosive interaction as the main process to be responsible for the formation of the crater in spite the fact that the flat-floored crater and its broad shape very common among tuff rings. The significance of this geosite is to demonstrate that explosion craters can be common features in association with lava dome-dominate eruptions. The easy access and the rewarding fantastic view can be used as main promotion features to attract visitors to this site.

In addition from this location the visitor can see the Mosawdah volcano and its extensive lava field just north of the Matan lava dome.

3.11.2 Mouteen Lava Dome Geotope [24° 12′51.79″N; 39° 50′38.82″E]

Along the 4WD road to the heart of the Precinct 3, just about 3 km from the Matan lava dome toward the SE, the visitor can view the Mouteen lava dome (Fig. 3.83). The lava dome seems to be smaller than the Matan lava dome with an about 800 m diameter. It also seems simpler than the Matan lava dome structure by having a simple flat toped lava dome (Fig. 3.84). The very unique nature of this lava dome however in comparison to the Matan lava dome is, that the silicic lava dome nearly completely intruded, invaded and truncated an earlier basaltic scoria cone that is exposed in the SE side of this volcanic complex.

3.11.2.1 Geosite—Mouteen Lava Dome Hosting Cinder Cone [24° 13′ 3.80″N; 39° 51′ 11.35″E]

From the 4WD road taking the visitor to the interior of the Precinct 3 a short walk the visitor can access the Mouteen lava dome. Prior reaching the lava dome itself, the walking track pass through the flank of a cinder cone that has been truncated and invaded by the lava dome. The cinder cone composed of black to red scoria that is littered by large spindle bombs especially in the region of the crater rim (Fig. 3.85). The contact between the lava dome and the cinder cone is covered by debris apron however it is clear that the more viscous lava dome truncated the host cinder cone. It is difficult to constrain what timeframe the lava dome extrusion took place, e.g. how long after the cinder cone formed the lava dome grown in its crater. However there are no strong evidences to support that the lava dome formation was immediately followed the scoria cone formation and it is more likely that there was significant time gap between these two strikingly different eruption style and magma composition to reach the surface. This geosite therefore has a great geoeducational value as it records rejuvenation of volcanism in the same place where previously different composition magma formed volcanoes through a very different eruption mechanism.

3.11.2.2 Geosite—Mouteen Lava Dome Main Dome Complex [24° 13′ 3.55″N; 39° 50′ 50.37″E]

This geosite can be accessed by further walking to the top of the Mouteen lava dome. The visitor can examine the lava dome material with great detail and in the end of the track

Fig. 3.83 GoogleEarth image of the Mouteen lava dome complex showing its complex geometry. The top left corner of the map view is [24° 14′ 5.32″N; 39° 49′ 29.02″E]

Fig. 3.84 Lava dome *top* of the Mouteen lava dome (*middle view*). Matan lava dome is visible in the *right* side of the view

Fig. 3.85 Spindle bombs form ballistic bomb region in the crater rim region of the cinder cone hosting the Mouteen lava dome [24° 13′ 3.80″N; 39° 51′ 11.35″E]

the top of the lava dome can be explored where spines, rock falls and inter-dome aeolian silt pans can be seen. From the end of the track in the North of the Mouteen lava dome top perfect view shows the Matan lava dome and in the distance the Mosawdah cone and lava field.

3.11.3 Jabal Al Malsaa Matam Volcanic Complex Geotope [24° 12′ 18.43″N; 39° 51′ 6.65″E]

Jabal al Malsaa Matam is a large volcanic complex with multiple vents and cone segments. From the top of Mouteen lava dome a perfect view shows its full extent as sitting on a large flat region covered by various pyroclastic density current deposits, lava flows, inter-cone stream valleys and accumulation zones of aeolian dust in lee sides of landforms (Fig. 3.86). The cone itself is large, and its volcanic architecture indicates that it must have been active long time, and it is likely represents a transitional volcano type between pure monogenetic and polygenetic volcanoes. This complexity of the volcano is apparent in its SW sector where various lava flows, craters, and pyroclastic rock units provide a complex set of volcanic edifice.

3.11.3.1 Geosite—Benmoreite Lava Flow [24° 11′ 35.78″N; 39° 51′ 21.37″E]

In the SW foothill of the volcano a spectacular benmoreite lava flow is exposed. The lava flow shows intensive flow banding and a very unique flow top morphology closely resembling clastic rock textures. In several places the flow is partially cross cut by wadis where the rock surface is more polished exposing the coherent texture of the rock. This geosite is important as it shows a graphic example to the visitor that the distinction between clastic and coherent rock textures is not easy when we deal with silicic rocks (McPhie et al. 1993). The disadvantage of this geosite is that it is difficult to access.

3.11.3.2 Geosite—Explosion Crater and View to the Main Cone [24° 11′ 41.26″N; 39° 51′ 21.13″E]

Just slightly above the benmoreite lava flow geosite, a low rimmed explosion crater exposes a thin pyroclastic succession indicative for magma and water explosive

Fig. 3.86 Jabal al Malsaa Matam volcanic complex is a volcano with multiple vents and compound volcanic edifice [24° 12′ 18.43″N; 39° 51′ 6.65″E]

interaction (Fig. 3.87). This crater is flat floored and filled nearly completely with voclaniclastic debris eroded from the southern flank of the southern cone of the Jabal al Malsaa Matam volcanic cone complex. The broad crater is partially surrounded by accidental lithic-rich silicic lapilli tuff and tuff beds. The significance of this geosite is that it provides some evidences to support that phreatomagmatism took place in the formation of this crater. In this respect this crater can be contrasted with that located in the west of the main Matan lava dome. It also highlights the fact that small and single silicic craters (e.g. just more west from the Matam lava dome complex) can likely be resulted from single and short-lived phreatomagmatic explosive eruptions similar to those reported elsewhere (Herrero-Hernandez et al. 2012; Németh et al. 2012; Zimmer et al. 2010).

Fig. 3.87 Exposed phreatomagmatic pyroclastic rock units surrounding a single explosion crater in the southern sector the Jabal al Malsaa Matam volcano [24° 11′ 39.75″N; 39° 51′ 28.15″E]

3.11.4 Um Junb Lava Dome Geotope [24° 11′ 59.43″N; 39° 53′ 27.63″E]

Um Junb is a lava dome complex with a 1.5 km diameter volcanic cone and lava dome structure (Fig. 3.88). The pyroclastic cone forming the base of the volcanic edifice has been invaded by benmoreitic magma that formed explosive and effusive eruptive products. The pyroclastic record of this volcano indicates an initial explosive eruption that produced pyroclastic flows that engulfed nearby obstacles, such as pre-existing scoria cone that lower flank is partially covered by the pyroclastic flow deposit. The pyroclastic flow deposits from Um Junb volcano are restricted and hardly traceable further than 500 m from the preserved, otherwise eroded volcanic edifice (Fig. 3.89). The volcano is only accessible by intensive walking. The geotope itself has significance that its crater has been filled by a lava dome, while the silicic tuff ring is still reasonable intact. In this respect the volcano represents a stage of volcano development where the subsequent explosive eruptions and intense lava dome emplacement has not reached that degree that the initial volcanic edifice suffered significant truncation of its geometry.

3.11.5 Dabaal Al Shamali Lava Dome Geotope [24° 13′ 20.02″N; 39° 54′ 19.07″E]

Dabaal al Shamali is a well-developed lava dome easy to access from a sealed road (Fig. 3.88). The lava dome has an asymmetric shape from side view (Fig. 3.90). The lava dome is surrounded by a rampart-like feature that is an earlier stage lava dome remnant truncated by the subsequent extrusion of further lava. The volcano southern side exposes large and deep fractures. These fractures are inferred to be related with the gravitational dilation caused by the extruding lava in the centre of the lava dome. While this part of the volcano is spectacular, and the cone flank in spite of its steep slope fairly stable it is not recommended to be visited alone and without proper gears due to potential collapse of some of the rim of these deep fissures. This geotope has a significant geoeducational value as it can show clearly that a lava dome can be gravitationally instable and such instability can cause a collapse (de Vries et al. 2014). It is also evident from this geotope that such collapse can take place accidentally during the lava dome growth or well after the growing ceased.

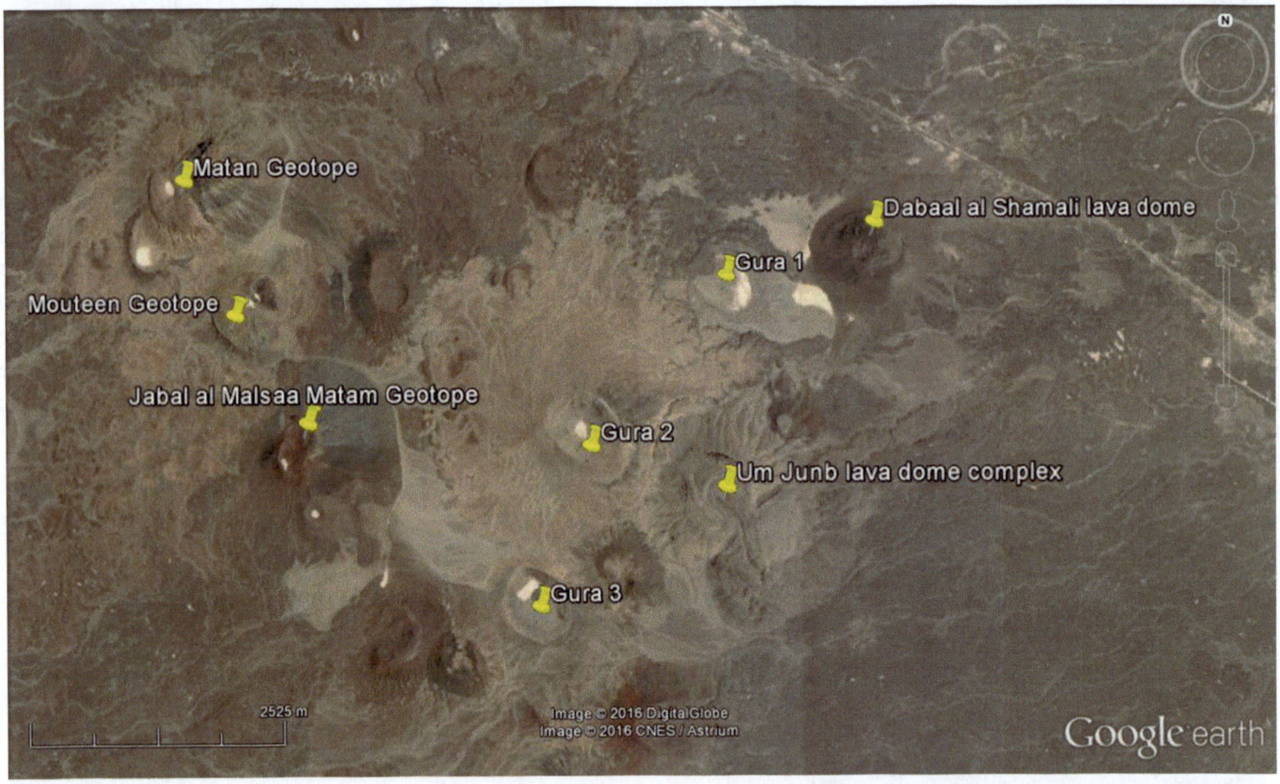

Fig. 3.88 Um Junb lava dome on GoogleEarth satellite image. Top left corner of the map view is [24° 14′ 40.01″N; 39° 49′ 6.05″E]

Fig. 3.89 Um Junb lava dome complex from about 3 km distance. Photo was taken from the point about [24° 13′ 20.22″N; 39° 52′ 42.24″E]

3.11.6 Gura 1 Explosion Crater Geotope [24° 13′ 5.58″N; 39° 53′ 29.36″E]

The Gura 1 volcano in the northern part of the Precinct 3 (Fig. 3.88) is a typical example of a "simple" explosion crater, which is a low-rimmed crater of 600 m diameter that has been partly overlapped by later a'a lava flows with abundant tumuli (Fig. 3.91). The crater rim of Gura 1 exposes pyroclastic surge-dominated sequences and poorly sorted explosion breccias that are dominated by a range of country rock lithologies. The deposits of the Gura 1 crater are confined to the rim around the crater and quickly pinch out over a few hundreds of metres from its crater rim. There are no widespread pyroclastic deposits (neither fall nor flow) associated with this crater, indicating that its formation was dominantly controlled by the explosive interaction of rising magma and ground water where the explosions generated ground-hugging base-surges that travelled a few hundreds of metres from their source (Austin-Erickson et al. 2011; Lorenz 2007; Németh and Cronin 2011). The basal deposits of the tuff ring are dominated by dense, angular trachytic lithologies, indicating the presence of silicic lava dome associated rocks beneath the vents that were disrupted by the explosive eruptions through the formation of Gura 1. Gura 1 can be interpreted as a shallow maar volcano, and highlights the potential effect of the presence of ground water on the resulting volcanic landform.

3.11.7 Gura 2 Explosion Crater Geotope [24° 12′ 11.68″N; 39° 52′ 42.02″E]

Gura 2 volcano is a large volcanic crater (Fig. 3.88) that has produced the one of the most widespread pyroclastic flow deposits in the area of the proposed Precinct 3 (Fig. 3.92). The widespread pyroclastic deposits cover the surrounding low lands and climb over small obstacles, and can be traced high up on distant and tall volcanic cones (Fig. 3.92). The pyroclastic flow deposits' light colour provides a dramatic landscape to the area that the visitor cannot miss. The pyroclastic flow-forming eruption style and the silicic (trachytic) composition of the magma involved in this eruption make Gura 2 volcano a unique place where a devastating and significant landscape-modifying volcanic eruption can be demonstrated to the visitors. This is a fundamental concept in the design of the geoeducational program of the proposed

Fig. 3.90 Dabaal al Shamali lava dome from the distance at a point of about [24° 12′ 6.21″N; 39° 54′ 58.36″E]

Fig. 3.91 Look out to the Gura 1 explosion crater from the distance looking toward Dabaal al Shamali lava dome. Photo was taken from a point about [24° 13′ 1.82″N; 39° 53′ 22.43″E]

Precinct The general preconception for the eruption style and eruption effect of a mafic intra-continental volcanic field is generally mildly explosive and largely effusive. While this is certainly the case for the youngest eruptions of the Harrat Rahat—presented in the area of Precinct 1 and 2—Gura 2 volcano provides a graphic example that highly destructive and explosive eruptions took place in the Harrat Al Madinah in the not too distant pass (about 0.7 Ma). Gura 2 is also a spectacular volcanic landform with enormous aesthetic and adventure volcanic tourism value. It consists of a crater of 500 m diameter within a tuff ring of 700–800 m diameter (Fig. 3.93). The inner crater wall at the eastern edge of the volcano exposes an earlier constructed evolved basaltic cone and lava flow complex, which was cut in half by the explosion and potential crater floor subsidence (Fig. 3.93). While the crater gives the impression that its floor is below the syn-eruptive landscape, and therefore it should be defined as a maar volcano, the reality is different. The crater floor is about 100 m above the base of the volcano and therefore the entire volcano forms a broad lensoid shape

Fig. 3.92 Gura 2 volcano produced block-and-ash flow deposits (light coloured cover over landscape) covered the older cones and inter-cone regions up to 10 km from the source. Photo was taken from the southern crater rim of Gura 2 [24° 12′ 16.42″N; 39° 52′ 31.50″E] toward south

Fig. 3.93 Crater of the Gura 2 volcano exposes older lava domes and half sectioned scoria cone segments suggesting complex eruption history of the volcano

positive landform. The tuff ring surrounds the main crater dissected the pre-tuff ring scora cone and lava dome forms a low angle outward dipping rim that is gradually transforming into a pyroclastic flow deposit-covered landscape mantling successions traceable over several kilometres from the crater.

The primary pyroclastic deposits further form reworked pyroclast-fans entering into neighbouring valleys, providing a valuable, perfectly exposed volcanic facies association traceable from the source to its reworked fans (Fig. 3.94). Gura 2 volcano in this respect gives an opportunity to study

Fig. 3.94 Gura 2 volcano forms a positive landform with a deep and wide crater on top. Block-and-ash flow deposits derived from Gura 2 volcano cover the landscape several kilometres away from the volcano (light cover over landscape). Note the crater of Gura 3 volcano in the *left* side of the view

the interaction between pyroclastic flows and pre-existing topography, as well as post-depositional reworking processes and therefore this geotope has a significant geoeducational value rarely visible such clearly elsewhere (Brown et al. 2003; Brown and Branney 2004; Edgar et al. 2007). The accumulated pyroclastic successions are perfectly exposed and the lack of vegetation cover makes it a perfect playground for experts and general public to investigate the effect and style of pyroclastic flow deposit accumulation.

3.11.7.1 Geosite—Western and Northern Crater Rim

The crater rim of Gura 2 volcano can be accessed from the south. To the southern crater rim a 4WD track can take the visitor then a westerly round trip can be taken by foot to the northern rim. On the way to the southern crater rim lookout the path follow a typical block-and-ash flow fan surface littered by moderately vesicular trachyte lapilli and block hosted in a fine ash matrix. This rock texture is evident in the crater rim where proximal sections of the block-and-ash flow deposits exposed. Toward the north it is clearly visible how the pyroclastic flow deposits mantle over the disrupted trachytic lava domes as an older volcanic landform dissected by the explosive eruptions of Gura 2. This geosite has a high geoeducational value as it demonstrates clearly how crater formation, pre-existing volcanic landforms and the depositing pyroclastic successions from the disrupting explosive events form together the volcanic landscape.

3.11.7.2 Geosite—Eastern Gully [24° 12′ 11.33″N; 39° 53′ 4.83″E]

Gura 2 volcano provides a more violent and complex eruption scenario in comparison to Gura 1 that can be closely examined in a deep incised gully network in the eastern margin about 1 km from the crater rim that can be defined as a key geosite. In this gully a at least three major pyroclastic flow unit is exposed in their proximal facies that initiated from the Gura 2 volcano suggesting major explosive eruption that give birth to this volcano (Fig. 3.95). It is inferred that the explosive eruptions were initiated by an explosive interaction of trachyte intrusion with groundwater, leaving behind the basal dense, green tuff breccia. The initial vent opening was quickly followed by eruptive phases producing pyroclastic flows that travelled up to several kilometres. The initial cratering and un-roofing that was likely triggered by phreatomagmatism was followed by volatile-rich trachyte magma emplacement, triggering a series of explosive eruptions, generating dense and particle-charged, “*heavy*” eruption clouds—probably about 3 to 10 km tall—that quickly collapsed and inundated an area of about 20 km^2 by pyroclastic flow and surge deposits.

3.11.8 Gura 3 Explosion Crater Geotope [24° 11′ 22.71″N; 39° 52′ 24.36″E]

Gura 3 volcano is another volcanic crater in the Precinct 3 (Fig. 3.88); it has a crater diameter of about 500 m (Fig. 3.96) and similar geometry and deposits as documented for Gura 1. The crater rim exposes pyroclastic surge deposits and consists of rare dense trachyte-dominated juvenile pyroclasts that are predominantly accidental lithics clasts of altered syenite and meta-sedimentary rocks. A small spine-like trachytic lava dome occupies the centre of the flat crater floor and indicates that in the late stage of the eruption, after the crater formed, some degassed melt was able to squeeze through (Fig. 3.97). The laterally restricted deposits of the Gura 3 volcano are covered by pyroclastic flow deposits derived from the Gura 2 volcano, providing a relative stratigraphic age; however there is no evidence of significant erosion and/or soil formation on top of the Gura 3 and below the Gura 2 deposits. Similarly to Gura 1, Gura 3

Fig. 3.95 Proximal pyroclastic flow units in the eastern sector of the Gura 2 volcano about 1 km from its crater rim [24° 12′ 11.33″N; 39° 53′ 4.83″E]

is also the result of an explosive interaction between rising trachytic magma and ground water. The volcano can be accessed directly by 4WD car, and a short walk can complete a perfect view of a well-developed maar volcano, that is unique feature in the Harrat Rahat and therefore it has a significant geoeducational value.

3.11.9 Al Shaatha Volcanic Complex Geotope [24° 8′ 39.75″N; 39° 53′ 35.62″E]

The Al Shaatha volcanic complex is a difficult to access region in the centre of the Harrat Al Madinah (Fig. 3.98). A poor quality 4WD dirt road need to be followed leaving the Gura 2–3 volcanoes that made a circle toward the southern edge of the Al Shaatha volcanic complex. It seems that the volcanic complex erupted through an older mafic (basaltic) volcanic chain and the trachytic eruption(s) partially destroyed the pre-trachyte cones leaving behind a moon-like landscape very difficult to travel and be oriented. The Al Shaatha volcanic complex seems to form a large broad crater that is filled with trachytic pyroclastic deposits that leap over the crater rim largely defined by the pre-trachyte volcanic morphology. Toward the south a major block-and-ash flow fan similar to those reported elsewhere (Freundt et al. 2000; Siebe et al. 1993) filling a flat floored wide valley, while toward the east fine grained pyroclastic debris covers a large area in the centre of the Harrat Al Madinah.

3.11.9.1 Geosite—Pyroclastic Flow Fan [24° 8′ 50.34″N; 39° 52′ 4.31″E]

This geosite is a unique place as it preserves a 500 m wide block-and-ash fan (Charbonnier and Gertisser 2008; Freundt et al. 2000) proximal to medial section that is gradually feeding into a normal volcaniclastic fan accumulating in a wadi (Fig. 3.98). The block-and-ash fan can be accessed easily from the dirt road and a short walk can take the visitor to the region where more than a metre across pumiceous trachyte fragments can be seen close (Fig. 3.99). From the access point the visitor can view the proximal areas of the block-and-ash fan as the deposits leap over the crater rim of

Fig. 3.96 View of the Gura 3 maar [24° 11′ 28.50″N; 39° 52′ 23.97″E] from the south showing its deep crater with a small toloid. View is toward NE

Fig. 3.97 Typical pyroclastic succession of the Gura 3 maar geotope's western crater rim and the small toloid in the middle of the crater. Picture was taken from [24° 11′ 30.52″N; 39° 52′ 14.02″E] toward SE

the volcano about 150 m above the wadi floor. The geosite has great geoeducational value as it demonstrate the valley-confined nature of the dense block-and-ash flows that can fill valleys (Gertisser et al. 2012; Schwarzkopf et al. 2005; Sulpizio et al. 2010; Suzuki-Kamata et al. 2009).

3.11.9.2 Geosite—Trachytic Ash Plain [24° 7′ 50.00″N; 39° 53′ 56.61″E]

In the eastern side of the Al Shaatha volcanic complex, the terrain is covered by light coloured, pumiceous deposit forming an extensive ash plain that shown up clearly on the

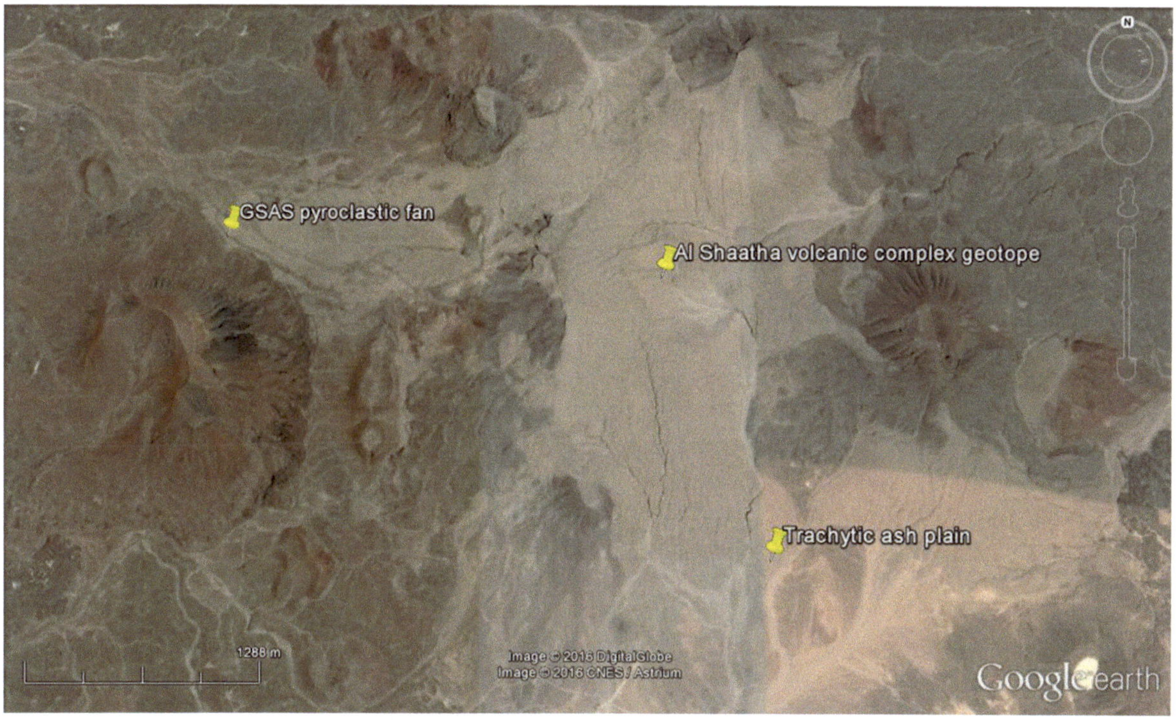

Fig. 3.98 GoogleEarth satellite image of the Al Shaatha volcanic complex geotope (GSAS—Geosite Al Shaatha). Top left corner of the map view is about [24° 9′ 37.86″N; 39° 51′ 8.83″E]. Note the white tone of the image representing trachytic ash plain and block-and-ash flow fans

Fig. 3.99 Large pumiceous trachyte block with radial jointing pattern typical for a block-and-ash flow that generated by a lava dome collapse and associated explosive event [24° 8′ 54.32″N; 39° 52′ 6.25″E]

satellite images (Fig. 3.98). The ash plain is soft and aeolian remobilisation of pyroclasts is intensive. This geosite has a high geoeducational value as it can demonstrate the potential aftermath of a silicic eruption that covers large regions with fine, light coloured pumiceous ash. The frequent sand storms and the aeolian remobilisation of this sediment can provide valuable ideas to scale the devastation a large silicic pyroclast-producing explosive eruption can cause through block-and-ash flow inundation and airfall (Major et al. 2013; Pittari et al. 2006; Vernet and Raynal 2008). Also, the silicic ash plain shows numerous sedimentological features to better understand the resetting of the landscape after a major silicic ash-producing eruptions through various fluvial and aeolian processes similarly to those regions commonly experienced recent silcic eruptions (Aceves Quesada et al. 2014; Cuitino and Scasso 2013; Umazano et al. 2014). In this respect this geosite provide a complementary site to the basaltic ash plain sites such as those documented near the 1256 AD historic eruption site (Kawabata et al. 2015) as part of the proposed Precinct 1.

3.11.10 Gura 4 Explosion Crater Geotope [24° 6′ 47.80″N; 39° 55′ 56.70″E]

Gura 4 explosion crater is a large and deep crater that can be accessed from the Al Shaatha volcanic complex by a truly adventure style trip. The Gura 4 volcano sits in a middle of a major block-and-ash fan region where trachytic block-and-ash flow deposits from various sources covered the region (Fig. 3.100). The crater is nearly 700 m wide and its shape is fairly irregular with some scalloped surfaces and rock falls (Fig. 3.101). In the crater wall red scoriaceous pyroclastic rock untis are exposed that are mantled by silicic tephra partially originated from the Gura 4 vent. The exact stratigraphy order of the pyroclastic rock units in this region is not established yet, and therefore it cannot be said more about the relative timing of events formed the major craters and major block-and-ash flow fans in this region. This geotope is significant not only by its spectacular landscape but also the deep crater in what the pre-crater scoria cones are half-sectioned indicating that pre-eruption volcanic morphology has some effect on the volcanic landform an explosive, crater-forming event can create.

3.11.11 Gura 5 Explosion Crater and Block-and-Ash Fan Geotope [24° 6′ 14.86″N; 39° 57′ 9.94″E]

Gura 5 is a relatively small crater which is located relatively far from other craters (Fig. 3.100). The crater is unique by its appearance as being separated from other craters as indicates that it might represent a single explosive event that occurred in this region without any major development of lava domes. The peculiarity of this volcano is that it forma a rather positive landform and therefore it differs from the Gura 1 single explosion crater where the crater is clearly wide, broad and the tephra ring surrounding it inferred to have formed by phreatomagmatic explosions. Gura 5 is inferred to be a positive landform and probably erupted through moderate explosive eruptions of trachytic magma. There is no clear evidence to support phreatomagmatism in the formation of this volcano. This region also hosts another unique geological feature that has great geoeducational significance. Just west of the Gura 5 crater a deep canyon exposes a thick (30 m +) complex pyroclastic flow successions clearly deposited from multiple eruption events, and potentially from multiple sources. The landscape forms a flat surface that cut through deep box canyon where the typical block-and-ash flow units are exposed. The landscape is very unique in volcanic fields and it is more common feature in areas where long-lived silicic volcanoes exist, and produced multiple pyroclastic flow (e.g. ignimbrite) sheets that covered the landscape such as those in Tenerife (Brown and Branney 2013; Bryan et al. 1998; Garcia et al. 2011; Smith and Kokelaar 2013), or Central Anatolia (Aydar 1998; Le Pennec et al. 2005). This well-preserved "*ignimbrite landscape*" in spite its remoteness can provide a very unique view on Harrat Al Madinah and therefore this geotope and its geosites bear a major geoeducational value. This geotope shows clearly that on a volcanic field that is largely composed of basaltic volcanoes that erupted small-volume eruptive products that resulted mostly scoria cones and lava flows, extensive sheet like ignimbrite eruptions can take occur. This has a fundamental geohazard message. This geotope can provide to the visitor key information that this geotope and all the nearby sites as part of Precinct 3 can offer a very important volcanic hazard education tool for the public to be utilized through volcanic geoheritage, geoconservation and geotourism.

3.11.12 Um Raqubah Lava Dome Geotope [24° 5′ 23.44″N; 39° 57′ 45.18″E]

Um Raqubah lava dome is a spectacular lava dome in the southern edge of Harrat Al Madinah (Fig. 3.100). It can be accessed from a sealed road that joins to a dirt road toward the interior of the Harrat Al Madinah's Precinct To visit this site well equipped 4WD car and good navigational gear and expertise are needed, as the road follows Bedouin hunting tracks winding around older aa lava fields. The lava dome is a major landmark as it can be visible from far by its typical telescope shape lava dome that has a major spine in its top (Fig. 3.102). The lava dome surrounded by a steep

Fig. 3.100 The Gura 4, 5, Um Rgaibah and Al Efairia crater in GoogleEarth satellite image. Top left corner of the map view is about [24° 7′ 8.28″N; 39° 53′ 21.40″E]

Fig. 3.101 The Gura 4 crater [24° 6′ 47.55″N; 39° 55′ 56.30″E] from the air showing that's the crater cut through pre-existing scoriaceous successions probably part of a chain of an older scoria cone system

Fig. 3.102 Um Raqubah lava dome with its telescopic lava dome complex and block-and-ash fan [24° 5′ 23.44″N; 39° 57′ 45.18″E]

block-and-ash fan that gradually diminishes in wadi networks about a km from the lava dome centre. Walking track can take the visitor to the first steep wall of the lava dome relatively easily by following the top of the block-and-ash fans. Further climb needs more care as the top spine is steep and cross cut by abundant fractures. This lava dome stands as a single lava dome with multiple growth stage in a volcanic field, and in this respect it is similar to lava dome-fields reported in intracontinental settings such as those in Anatolia (Seghedi et al. 2015; Sen et al. 2002; Siebel et al. 2011). The significance of this geotope is that it provides a very nice view of the architecture of a lava dome formed by multiple intrusive stages and also fed some block and ash flows that deposited their block-and-ash fans around the growing lava dome. This gradual step-by-step growth of the dome and the evolution of the block-and ash fan is clearly visible in this region. In the southern margin of the lava dome a deep gully network exposes a fantastic set of block-and-ash flow deposits inferred to be derived from the nearby Al Efairia volcanic complex. While to visit this geosite it is not important from where the block-and-ash flow came from, the significance of this site is the perfect cross-sectional view of a block-and-ash flow fan where block- and pumice-rich part of the units are well exposed (Fig. 3.103). The top of the block-and-ash fan are typically littered by large convolute and rugged blocks of trachyte mimicking the surface of a lava flow (Fig. 3.104). The significance of this geosite is to demonstrate the typical surface morphology of a block-and-ash fan typical to the Precinct 3 in the proposed Harrat Al Madinah Volcanic Geopark.

3.11.13 Al Efairia Volcanic Complex Geotope [24° 4′ 29.28″N; 39° 56′ 19.40″E]

Al Efairia is a large volcanic complex in the southern part of the Precinct 3 of the proposed HArrat Al Madinah Voclanic Geopark (Fig. 3.100). This complex volcano is best to be defined as a large silicic eruptive center that likely produced some shallow volcanic crater and multiple lava dome complexes. It closely resembling many complex lava domes sitting in silicic tuff rings in dome-tuff ring fields commonly reported from the geological record in various geotectonic settings (Brooker et al. 1993; Henry et al. 1997; Lexa et al. 2010; Riggs et al. 1997). The actual crater (or main eruptive vent) of the volcanic complex is difficult to reconstruct as

Fig. 3.103 Block-and-ash flow units derived from the Al Efairia volcanic complex showing graphic example of the internal architecture of a block and ash flow fan common in the region of the Precinct 3 (and 4) [24° 4′ 38.70″N; 39° 57′ 52.93″E]

Fig. 3.104 Surface of a block-and-ash fan littered by variously vesicular trachytic clasts closely resembling fragments of lava flow surfaces. The slopes are also difficult to path by 4WD or even by walking. Note Um Raqubah lava dome in the back

just part of one of the crater is preserved well. The volcano also complicated by the fact that it has been erupted through a pre-existing scoria cone dominated volcanic chain that has been partially dissected by the formation of the Al Efairia volcano. It is also evident that the volcano had multiple vents and vent shifting produced some slightly migrated vent setting along lava domes formed. While the volcano proximal area is complicated and rather resembles a compound lava dome-dominated silicic volcano, the extensive block-and-ash flow fans are relatively simple and covering probably the largest surface area of the Harrat Al Madinah's silicic volcanoes. This volcanic geotope has huge geoeducational significance as it provides the largest and most explosive style eruption scenario any volcanic hazard study needs to deal with in future eruptions in the Harrat Al Madinah region.

3.11.13.1 Geosite—Caldera View [24° 4′ 35.58″N; 39° 56′ 56.45″E]

From the NE edge of the crater rim a view can be seen toward the crater (small caldera) of the Al Efairia volcanic complex (Fig. 3.105). The crater is slightly east-west elongated and filled with aeolian dust as reworked pumiceous ash from the nearby ash plains. The crater rim is clearly formed by block-and-ash flow deposits that mantle the landscape toward the north. The geosite is an important lookout point as from here the block-and-ash flow run out distances can be seen clearly, and that information can provide important data for eruption scenario studies for future volcanic eruptions. From this look out it is also evident that the southern side of the volcano is truncated by lava dome complexes, while the northern crater rim partially covers disrupted scoria cones.

3.11.13.2 Geosite—Half-Sectioned Pre-caldera Cone [24° 4′ 54.86″N; 39° 56′ 35.69″E]

In the northern crater rim half section of a scoria cone core exposed (Fig. 3.106). The exposed scoria cone composed of welded and agglutinated basaltic scoria. The cone itself is covered by light coloured pumiceopus block-and-ash flow deposits of the Al Efairia volcano. The block-and-ash flow deposits cover nearly the entire pre-crater scoria cone. Along an east—west line several scoria cones are half-sectioned and partially covered by block-and-ash flow deposits inferred to be derived from the Al Efairia cone. This geosite is probably the best geosite in the entire proposed Al Madinah Volcanic Geopark, where the rejuvenation of silicic volcanism along a pre-existing mafic volcanic chain is clearly visible. This volcanic facies architecture has a major significant on the melt source stability and maybe some structural influence on the rejuvenation of volcanism. Therefore this geosite has a fundamental significance for geoeducation program the proposed geopark intend to promote (Fig. 3.107).

3.11.13.3 Geosite—Al Efairia Lava Dome Complex [24° 4′ 29.28″N; 39° 56′ 19.40″E]

The lava domes of the Al Efairia volcanic complex are not as spectacular visually than optehr previously described lava domes, however they bear geological significance as they seem to grown into a broad crater formed the major and extensive block-and-ash fans derived from the AL Efairia volcanic complex. The lava domes expose flow banded trachyte that is partially flanked by short run-out distance block-and-ash flow deposits with coarse units. The geoeducational value of this geosite is the potential of the clear demonstration to the link between explosive and effusive phase of a silicic volcanic complex in the region. Similar relationships have been demonstrated elsewhere in the Precinct 3, however this site provides the most complex scenario and the largest volume of deposits.

3.11.13.4 Geosite—Complex Volcaniclastic Succession of a Block-and-Ash Fan [24° 4′ 8.49″N; 39° 56′ 38.49″E]

In the SW edge of the Al Efairia volcanic complex a well-developed block-and-ash fan is exposed. The block-and-ash fan shows perfectly the intermixing of primary and secondary deposits on this exposed gentle sloping region. The geosite has a geoeducational significance to demonstrate that in an active inter-con/inter-dome region primary and secondary volcanic processes together forming the landscape and the depositional environment. Moreover

Fig. 3.105 View of the main crater of the Al Efairia volcanic complex. Photo was taken from [24° 4′ 35.58″N; 39° 56′ 56.45″E] toward W

Fig. 3.106 Half section of a proximal scoria cone in the northern crater rim of the Al Efairia volcanic complex [24° 4′ 54.86″N; 39° 56′ 35.69″E]

Fig. 3.107 Well-developed block-and-ash fan in the SW margin of the Al Efairia volcano. In the background the large trachytic lava domes are apparent. Photo was taken from [24° 4′ 47.68″N; 39° 56′ 0.42″E] toward NW

from this geosite the crater filling lava domes are clearly visible with their block-and-ash fan also fed material to the distal silicic ash plain.

3.11.14 Al Wabarah Volcanic Complex Geotope (Precinct 4) [24° 0′ 52.87″N; 39° 53′ 16.22″E]

About 5 km south from Al Efairia is the very remote Al Wabarah volcanic complex. This region doesn't provide extra geological information to understand the volcanism of the Harrat Al Madinah, but by its remote location and rugged, untouched nature can be promoted as an alternative small scale version of Precinct 3, and could be defined as Precinct 4. The remoteness of the region ensures that the landscape has no human influence (yet) and can be promoted as a destination of ecotourism especially with some combination of promoting value of botany and zoology. The region composed of scoria cones similar to those detailed in the Precinct 1, and between these older scoria cone fields there is a double volcanic edifice with two broad craters that are seemingly the source of a succession of silicic block-and-ash flow deposits (Fig. 3.108). The region exposes block-and-ash fans where the visitor can see clear transition between primary and reworked volcaniclastic deposits to demonstrate the terrestrial sedimentary environment influenced by silicic explosive eruptions.

3.12 Geopark Potential of the Harrat Al Madinah—A Discussion

Organisation of a geopark along volcanic features with high geoheritage value is the way to maximaze the potential of a region to be promoted successfully as a geopark. In this chapter it has been demonstrated that a systematic arrangement of volcanic geoheritage value of the Harrat Al Madinah has a great potential to provide firm scientific basis to develop and function a geopark in the region. The presented precinct concept showed a logical approach to demonstrate the volcanic geoheritage of the region which is by surface area very large, and by its logistical aspects need to be designed in a way that visitors can cover areas that present logically set geoheritage sites that all together can provide extra added value to understand not only the volcanism behind the formation of the Harrat Al Madinah, but also its volcanic hazards the local population faces with. The main aim to develop a volcanic geopark in the youngest volcanism in the Kingdom of Saudi Arabia is to show the fundamental geological features a typical harrat can have. Harrat Al Madinah with its perfect logistical background and

Fig. 3.108 GoogleEarth satellite image of the Al Wabarah volcanic region as the provisional location of Precinct 4 [24° 0′ 52.87″N; 39° 53′ 16.22″E]

proximity to a cultural focal point on Earth that act as a magnet of large number of visitors is a logical start point to develop projects that promote a geological feature in the Arabian peninsula, which is a very common landscape in the western Arabian region and has numerous cultural implications as well. We demonstrated that potential geoeducational routes and trails across Harrat Rahat, particularly in its northern sector, the Harrat Al Madinah can offer endless world-class geosites that can serve significant knowledge transfer on volcanic landscape evolution and understanding the volcanic hazards a young harrat may indicate. It has been demonstrates that some suggested visitor itineraries through a carefully design geological precincts and their geotopes and geosite could be logically linked together to demonstrate specific volcanic features or processes associated with dispersed volcanism such as formation of intracontinental volcanic fields. The international volcanological significance of the Harrat Al Madinah especially its potential link to volcanic researches on intracontinental volcanic fields hold the potential of its international linkages to similar trails with other volcanic regions' geopark programs. The proposal and establishment of the HAMVG is the first such attempt in the territory of the Kingdom of Saudi Arabia. The success of this project will likely affect future geoconservation and geoeducation projects planned elsewhere in the Kingdom especially in other harrats. In the next chapters a brief summary will be given about the geoheritage and geoeducational potential of other, less known harrats of the Kingdom. In the following chapters we will not provide such detailed geosite level description of the geoheritage value of each of the harrats as such work would be way to extensive, but will provide significant link to Harrat Al Madinah, and will demonstrate the potential to develop a national network of volcanic geoparks that all together could even provide firm basis to apply candidature for being listed as a world heritage site as a fine example of dispersed intracontinental volcanism on Earth. The success of the Harrat Al Madinah Volcanic geopark potentially could provide an example for future similar activities elsewhere in the Arabian Peninsula and it could be designed as a flagship projects in the region. The impact of the proposed geopark on geotourism is expected to be huge. The Harrat Al Madinah's volcanic geology is the perfect place to see a wide array of volcanic features, dramatic volcanic landscapes and the interaction between the extreme climate and volcanic landforms associated with intra-continental monogenetic volcanism. The Harrat Al Madinah is a globally unique intra-continental monogenetic volcanic field due to (1) the large number of young (<1 Ma) monogenetic volcanoes it hosts, (2) the wide range of chemical compositions of magma that formed the specific volcano types (from basaltic to trachytic), (3) the diverse eruption styles (e.g. from Hawaiian-style and Strombolian-style magmatic explosive eruptions to phreatomagmatic maar and/or tuff ring forming eruptions) and (4) the dramatically different volcanic landforms and associated volcanic rocks (e.g. from lava spatter cones and scoria cones to trachytic maar volcanoes and lava domes) that can be seen in a perfectly exposed manner. The diversity of volcano types in the proposed geopark that can be visited and seen perfectly is probably among the greatest in comparison to other intra-continental monogenetic volcanic fields. It is suggested that the proposed geopark is linked informally with geoeducation, geoconservational and geotouristic activities conducted and promoted by potential "*sister geoparks*" elsewhere, such as the Bakony- Balaton Geopark (Hungary), Nógrád/Novohrad Geopark (Slovakia/Hungary), Vulkaneifel Geopark (Germany), Jeju Island Geopark (Korea), Unzen Volcanic Area Geopark (Japan), Kanawinka Geopark (Australia) and Wudalianchi Geopark (China). These "*sister geoparks*" demonstrate volcanic features associated with similar types of volcanism to the Harrat Al Madinah, but with differences that make the parks complementary, including the level of vegetation cover (e.g. vegetation covered maars with lakes from the Vulkaneifel Geopark that are in contrast with the Harrat Al Madinah's dry maars and explosion craters), exposure level (e.g. exposed volcanic conduits of similar monogenetic volcanoes to those at the Harrat Al Madinah are visible from the Bakony- Balaton and Nógrád/Novohrad Geopark), and variations in volcano types in accordance with variations in the volcanic eruption styles that formed them (e.g. active lava domes from Unzen Volcanic Area Geopark or the great size, shape and volcanic succession variations in Jeju Island, Korea). Linking and partially coordinating the scientific research, geoeducational/geoconservation programs and the geotouristic aspects of these "*sister geoparks*" would be a desirable approach in the future.

References

Abdurachman EK, Bourdier JL, Voight B (2000) Nuees ardentes of 22 November 1994 at Merapi volcano, Java, Indonesia. J Volcanol Geoth Res 100(1–4):345–361

Aceves Quesada F, Legorreta Paulin G, Alvarez Ruiz Y (2014) Geomorphologic mapping for the inventory of gravitational processes in the endorrheic basin of the La Cienega gulch, Eastern flank of the Nevado de Toluca. Boletín de la Sociedad Geológica Mexicana 66(2):329–342

Alvarado GE, Perez W, Vogel TA, Groeger H, Patino L (2011) The Cerro Chopo basaltic cone (Costa Rica); an unusual completely reversed graded pyroclastic cone with abundant low vesiculated cannonball juvenile fragments. J Volcanol Geoth Res 201(1–4):163–177

Ambraseys NN, Melville CP, Adams RD (1994) The seismicity of Egypt, Arabia and the red Sea: a historical review. Cambridge University Press, Cambridge, UK

Anderson SW, Fink JH, Rose WI (1995) Mount St Helens and Santiaguito lava domes: The effect of short-term eruption rate on surface texture and degassing processes. J Volcanol Geoth Res 69 (1–2):105–116

Anderson SW, Smrekar SE, Stofan ER (2012) Tumulus development on lava flows: insights from observations of active tumuli and analysis of formation models. Bull Volcanol 74(4):931–946

Anderson SW, Stofan ER, Smrekar SE, Guest JE, Wood B (1999) Pulsed inflation of pahoehoe lava flows: implications for flood basalt emplacement. Earth Planet Sci Lett 168(1–2):7–18

Armiero V, Petrosino P, Lirer L, Alberico I (2011) The GeoCaF project: proposal of a geosites network at Campi Flegrei (Southern Italy). Geoheritage 3:195–219

Austin-Erickson A, Ort MH, Carrasco-Nunez G (2011) Rhyolitic phreatomagmatism explored: Tepexitl tuff ring (Eastern Mexican Volcanic Belt). J Volcanol Geoth Res 201(1–4):325–341

Aydar E (1998) Early Miocene to Quaternary evolution of volcanism and the basin formation in Western Anatolia: a review. J Volcanol Geoth Res 85(1–4):69–82

Baer G, Hamiel Y (2010) Form and growth of an embryonic continental rift: InSAR observations and modelling of the 2009 western Arabia rifting episode. Geophys J Int 182(1):155–167

Ball M, Pinkerton H, Harris AJL (2008) Surface cooling, advection and the development of different surface textures on active lavas on Kilauea, Hawai'i. J Volcanol Geoth Res 173(1–2):148–156

Bednarz U, Schmincke H-U (1990) Evolution of the Quarternary melilite-nephelinite Herchenberg volcano (East Eifel). Bull Volcanol 52:426–444

Bilotta G, Cappello A, Herault A, Vicari A, Russo G, Del Negro C (2012) Sensitivity analysis of the MAGFLOW Cellular Automaton model for lava flow simulation. Environ Model Softw 35:122–131

Blatter DL, Carmichael ISE, Deino AL, Renne PR (2001) Neogene volcanism at the front of the central Mexican volcanic belt: Basaltic andesites to dacites, with contemporaneous shoshonites and high-TiO2 lava. Geol Soc Am Bull 113(10):1324–1342

Bondre NR, Hart WK (2008) Morphological and textural diversity of the Steens Basalt lava flows, Southeastern Oregon, USA: implications for emplacement style and nature of eruptive episodes. Bull Volcanol 70(8):999–1019

Bourdier JL, Abdurachman EK (2001) Decoupling of small-volume pyroclastic flows and related hazards at Merapi volcano, Indonesia. Bull Volcanol 63(5):309–325

Brenna M, Cronin SJ, Smith IEM, Maas R, Sohn YK (2012) How small-volume Basaltic Magmatic systems develop: a case study from the Jeju Island volcanic field, Korea. J Petrol 53(5):985–1018

Bretar F, Arab-Sedze M, Champion J, Pierrot-Deseilligny M, Heggy E, Jacquemoud S (2013) An advanced photogrammetric method to measure surface roughness: application to volcanic terrains in the Piton de la Fournaise, Reunion Island. Remote Sens Environ 135:1–11

Brooker MR, Houghton BF, Wilson CJN, Gamble JA (1993) Pyroclastic phases of a rhyolitic dome-building eruption—pUketarata tuff ring, Taupo Volcanic Zone, New Zealand. Bull Volcanol 55(6):395–406

Brown RJ, Branney MJ (2004) Event-stratigraphy of a caldera-forming ignimbrite eruption on Tenerife: the 273 ka Poris Formation. Bull Volcanol 66:392–416

Brown RJ, Branney MJ (2013) Internal flow variations and diachronous sedimentation within extensive, sustained, density-stratified pyroclastic density currents flowing down gentle slopes, as revealed by the internal architectures of ignimbrites on Tenerife. Bull Volcanol 75(7):1–24

Brown RJ, Barry TL, Branney MJ, Pringle MS, Bryan SE (2003) The Quaternary pyroclastic succession of southeast Tenerife, Canary Islands: explosive eruptions, related caldera subsidence, and sector collapse. Geol Mag 140(3):265–288

Bryan SE, Cas RAF, Marti J (1998) Lithic breccias in intermediate volume phonolitic ignimbrites, Tenerife (Canary Islands): constraints on pyroclastic flow depositional processes. J Volcanol Geoth Res 81(3–4):269–296

Camp VE, Roobol MJ (1989) The Arabian continental alkali basalt province; Part I, evolution of Harrat Rahat, Kingdom of Saudi Arabia; with Suppl. Data 89-04. Geol Soc Am Bull 101(1):71–95

Camp VE, Hooper PR, Roobol MJ, White DL (1987) The Madinah eruption, Saudi Arabia: Magma mixing and simultaneous extrusion of three basaltic chemical types. Bull Volcanol 49(2):489–508

Camp VE, Roobol MJ, Hooper PR (1991) The Arabian continental alkali basalt province; Part II, evolution of harrats Khaybar, Ithnayn, and Kura, Kingdom of Saudi Arabia; with Suppl. Data 91-06. Geol Soc Am Bull 103(3):363–391

Cappello A, Vicari A, Del Negro C (2011) Assessment and modeling of lava flow hazard on Mt. Etna volcano. Bollettino Di Geofisica Teorica Ed Applicata 52(2):299–308

Carn SA, Watts RB, Thompson G, Norton GE (2004) Anatomy of a lava dome collapse: the 20 March 2000 event at Soufriere Hills Volcano, Montserrat. J Volcanol Geoth Res 131(3–4):241–264

Carracedo Sanchez M, Sarrionandia F, Gil Ibarguchi JI (2014) Post-depositional intrusion and extrusion through a scoria and spatter cone of fountain-fed nephelinite lavas (Las Herrerias volcano, Calatrava, Spain). Bull Volcanol 76(9):1–17

Carter A, de Vries BV, Kelfoun K, Bachelery P, Briole P (2007) Pits, rifts and slumps: the summit structure of Piton de la Fournaise. Bull Volc 69(7):741–756

Charbonnier SJ, Gertisser R (2008) Field observations and surface characteristics of pristine block-and-ash flow deposits from the 2006 eruption of Merapi Volcano, Java, Indonesia. J Volcanol Geoth Res 177(4):971–982

Coleman RC, Gregory R (1983) Cenozoic volcanic rocks. US Geological Survey Professional Paper, 287–287

Costa A, Wadge G, Melnik O (2012) Cyclic extrusion of a lava dome based on a stick-slip mechanism. Earth Planet Sci Lett 337:39–46

Crown DA, Baloga SM (1999) Pahoehoe toe dimensions, morphology, and branching relationships at Mauna Ulu, Kilauea Volcano, Hawai'i. Bull Volcanol 61(5):288–305

Cuitino JI, Scasso RA (2013) Reworked pyroclastic beds in the early Miocene of Patagonia: reaction in response to high sediment supply during explosive volcanic events. Sediment Geol 289:194–209

de Vries BvW, Marquez A, Herrera R, Granja Bruna JL, Llanes P, Delcamp A (2014) Craters of elevation revisited: forced-folds, bulging and uplift of volcanoes. Bull Volcanol 76(11)

Del Negro C, Fortuna L, Herault A, Vicari A (2008) Simulations of the 2004 lava flow at Etna volcano using the magflow cellular automata model. Bull Volcanol 70(7):805–812

Deraman A, Yahaya JH, Salim J, Idris S, Komoo I, Leman MS, Unjah T (2010) A geodiversity repository system based on digital object modeling: an applied research on langkawi geoparks. In: Soliman KS (ed) Knowledge management and innovation: a business competitive edge perspective vols 1, pp 1208–1217

Doniz-Paez J (2015) Volcanic geomorphological classification of the cinder cones of Tenerife (Canary Islands, Spain). Geomorphology 228:432–447

Doniz J, Romero C, Coello E, Guillen C, Sanchez N, Garcia-Cacho L, Garcia A (2008) Morphological and statistical characterisation of recent mafic volcanism on Tenerife (Canary Islands, Spain). J Volcanol Geotherm Res 173(3-4):185–195

Dowling R (2011) Geotourism's global growth. Geoheritage 3(1):1–13

Duncan RA, Al-Amri AM (2013) Timing and composition of volcanic activity at Harrat Lunayyir, western Saudi Arabia. J Volcanol Geoth Res 260:103–116

Duraiswami RA, Dole G, Bondre N (2003) Slabby pahoehoe from the western Deccan volcanic province: evidence for incipient pahoehoe-aa transitions. J Volcanol Geoth Res 121(3–4):195–217

Duraiswami RA, Gadpallu P, Shaikh TN, Cardin N (2014) Pahoehoe-a'a transitions in the lava flow fields of the western Deccan Traps, India-implications for emplacement dynamics, flood basalt architecture and volcanic stratigraphy. J Asian Earth Sci 84:146–166

Edgar CJ, Wolff JA, Olin PH, Nichols HJ, Pittari A, Cas RAF, Reiners PW, Spell TL, Marti J (2007) The late quaternary Diego Hernandez Formation, Tenerife: Volcanology of a complex cycle of voluminous explosive phonolitic eruptions. J Volcanol Geoth Res 160(1–2):59–85

El Difrawy MA, Runge MG, Moufti MR, Cronin SJ, Bebbington M (2013) A first hazard analysis of the Quaternary Harrat Al-Madinah volcanic field, Saudi Arabia. J Volcanol Geoth Res 267:39–46

Erfurt-Cooper P (2011) Geotourism in volcanic and geothermal environments: playing with fire? Geoheritage 3(3):187–193

Erikstad L (2013) Geoheritage and geodiversity management—the questions for tomorrow. In: Proceedings of the geologists association 124(4):713–719 [http://dx.doi.org/10.1016/j.pgeola.2011007.1003]

Farsani NT, Coelho C, Costa C (2011) Geotourism and geoparks as novel strategies for socio-economic development in rural areas. Int J Tour Res 13(1):68–81

Fassoulas C, Mouriki D, Dimitriou-Nikolakis P, Iliopoulos G (2012) Quantitative assessment of geotopes as an effective tool for geoheritage management. Geoheritage 4(3):177–193

Favalli M, Karatson D, Mazzarini F, Pareschi MT, Boschi E (2009) Morphometry of scoria cones located on a volcano flank: a case study from Mt. Etna (Italy), based on high-resolution LiDAR data. J Volcanol Geoth Res 186(3–4):320–330

Fink JH, Griffiths RW (1998) Morphology, eruption rates, and rheology of lava domes: insights from laboratory models. J Geophys Res Solid Earth 103(B1):527–545

Fornaciai A, Favalli M, Karatson D, Tarquini S, Boschi E (2012) Morphometry of scoria cones, and their relation to geodynamic setting: A DEM-based analysis. J Volcanol Geoth Res 217:56–72

Francis PW (1973) Cannonball bombs—new kind of volcanic bomb from Pacaya volcano, Guatemala. Geol Soc Am Bull 84(8):2791–2793

Freundt A, Wilson CJN, Carey SN (2000) Ignimbrites and block-and-ash flow deposits. In: Sigurdsson H, Houghton BF, McNutt SR, Rymer H, Stix J (eds) Encyclopedia of Volcanoes. Academic Press, San Diego, pp 581–600

Fuertes-Gutierrez I, Fernandez-Martinez E (2012) Mapping geosites for geoheritage management: a methodological proposal for the Regional Park of Picos de Europa (Leon, Spain). Environ Manage 50(5):789–806

Fujii T, Nakada S (1999) The 15 September 1991 pyroclastic flows at Unzen Volcano (Japan), a flow model for associated ash-cloud surges. J Volcanol Geoth Res 89(1–4):159–172

Gao WY, Li JH, Mao X, Zhang TR (2010) Discussion on genetic mechanism of hornitos in Wudalianchi volcanic field. Acta Petrologica Sinica 26(1):309–317

Gao W, Li J, Mao X, Li H (2013) Geological and geomorphological value of the Monogenetic volcanoes in Wudalianchi National Park, NE China. Geoheritage 5(2):73–85

Garcia O, Marti J, Aguirre G, Geyer A, Iribarren I (2011) Pyroclastic density currents from Teide-Pico Viejo (Tenerife, Canary Islands): implications for hazard assessment. Terra Nova 23(3):220–224

Gertisser R, Cassidy NJ, Charbonnier SJ, Nuzzo L, Preece K (2012) Overbank block-and-ash flow deposits and the impact of

valley-derived, unconfined flows on populated areas at Merapi volcano, Java, Indonesia. Nat Hazards 60(2):623–648
Gordon JE (2012) Rediscovering a sense of wonder: geoheritage, geotourism and cultural landscape experiences. Geoheritage 4 (1):65–77
Guest JE, Duncan AM, Stofan ER, Anderson SW (2012) Effect of slope on development of pahoehoe flow fields: evidence from Mount Etna. J Volcanol Geoth Res 219:52–62
Guilbaud M-N, Siebe C, Layer P, Salinas S (2012) Reconstruction of the volcanic history of the Tacambaro-Puruaran area (Michoacan, Mexico) reveals high frequency of Holocene monogenetic eruptions. Bull Volc 74(5):1187–1211
Hansen SE, DeShon HR, Moore-Driskell MM, Al-Amri AMS (2013) Investigating the P wave velocity structure beneath Harrat Lunayyir, Northwestern Saudi Arabia, using double-difference tomography and earthquakes from the 2009 seismic swarm. J Geophys Res Solid Earth 118(9):4814–4826
Harris AJL (2009) The pit-craters and pit-crater-filling lavas of Masaya volcano. Bull Volc 71(5):541–558
Harris AJL, Favalli M, Mazzarini F, Hamilton CW (2009) Construction dynamics of a lava channel. Bull Volc 71(4):459–474
Hasenaka T (1994) Size, distribution, and magma output rate for shield volcanos of the Michoacan-Guanajuato volcanic field, Central Mexico. J Volcanol Geoth Res 63(1–2):13–31
Hasenaka T, Carmichael ISE (1985) A compilation of location, size, and geomorphological parameters of volcanoes of the Michoacan-Guanajuato volcanic field, central Mexico. Geofis Int 24(4):577–608
Head JW, Wilson L (1989) Basaltic pyroclastic eruptions: influence of gas release patterns and volume fluxes on fountain structure, and the formation of cinder cones, spatter cones, rootless flows, lava ponds and lava flows. J Volcanol Geoth Res 37:261–271
Henriques MH, dos Reis RP, Brilha J, Mota T (2011) Geoconservation as an emerging geoscience. Geoheritage, 1–12
Henry CD, Kunk MJ, Muehlberger WR, McIntosh WC (1997) Igneous evolution of a complex laccolith-caldera, the Solitario, Trans-Pecos Texas: implications for calderas and subjacent plutons. Geol Soc Am Bull 109(8):1036–1054
Herault A, Vicari A, Ciraudo A, Del Negro C (2009) Forecasting lava flow hazards during the 2006 Etna eruption: using the MAGFLOW cellular automata model. Comput Geosci 35(5):1050–1060
Herrero-Hernandez A, Lopez-Moro FJ, Gomez-Fernandez F, Martin-Serrano A (2012) Interaction between intra-continental sedimentary basins and small-volume monogenetic volcanism: Argamasilla and Calzada-Moral basins, Campo de Calatrava Volcanic Field, Spain. J Iberian Geol 38(2):407–428
Hill BE, Connor CB, Jarzemba MS, La Femina PC, Navarro M, Strauch W (1998) 1995 eruptions of Cerro Negro volcano, Nicaragua, and risk assessment for future eruptions. Geol Soc Am Bull 110(10):1231–1241
Hintz AR, Valentine GA (2012) Complex plumbing of monogenetic scoria cones: new insights from the Lunar Crater volcanic field (Nevada, USA). J Volcanol Geoth Res 239:19–32
Hoblitt RP, Orr TR, Heliker C, Denlinger RP, Hon K, Cervelli PF (2012) Inflation rates, rifts, and bands in a pahoehoe sheet flow. Geosphere 8(1):179–195
Holcomb RT (1981) Kilauea Volcano, Hawaii; chronology and morphology of the surficial lava flows. Open-file report—US Geological Survey, 335–335
Hooper DM, Sheridan MF (1998) Computer-simulation models of scoria cone degradation. J Volcanol Geotherm Res 83:241–267
Ilyinskaya E, Martin RS, Oppenheimer C (2012) Aerosol formation in basaltic lava fountaining: Eyjafjallajokull volcano, Iceland. J Geophys Res Atmos 117
Inbar M, Risso C (2001) A morphological and morphometric analysis of a high density cinder cone volcanic field—Payun Matru, south-central Andes, Argentina. Zeitschrift für Geomorphologie 45(3): 321–343
Inbar M, Gilichinsky M, Melekestsev I, Melnikov D, Zaretskaya N (2011) Morphometric and morphological development of Holocene cinder cones: a field and remote sensing study in the Tolbachik volcanic field, Kamchatka. J Volcanol Geoth Res 201(1–4):301–311
Jonsson S, Pallister J, McCausland W, El-Hadidy S, Anonymous (2010) Dyke intrusion and arrest in Harrat Lunayyir, western Saudi Arabia, in April-July 2009. Geophysical Research Abstracts 12: EGU2010-7704
Joyce E (2010) Australia's geoheritage: history of study, a new inventory of geosites and applications to geotourism and geoparks. Geoheritage 2(1–2):39–56
Juynboll GH (1989a) The History of al-Tabari Vol. 13: The Conquest of Iraq, Southwestern Persia, and Egypt: The Middle Years of'Umar's Caliphate AD 636–642/AH 15-21. State University of New York, USA, New York, USA, p 251
Juynboll GHA (1989b) Between fact and fiction, a study on the Umayyad Caliph Ibnabdalmalik, Sulayman and his image in the sources—German—Eisener R. Islam-Zeitschrift Fur Geschichte Und Kultur Des Islamischen Orients 66(2):358–361
Kaneko T, Wooster MJ, Nakada S (2002) Exogenous and endogenous growth of the Unzen lava dome examined by satellite infrared image analysis. J Volcanol Geoth Res 116(1–2):151–160
Kawabata E, Cronin SJ, Bebbington MS, Moufti MRH, El-Masry N, Wang T (2015) Identifying multiple eruption phases from a compound tephra blanket: an example of the AD1256 Al-Madinah eruption, Saudi Arabia. Bull Volcanol 77(1):1–13
Kazancı N (2012) Geological background and three vulnerable geosites of the Kızılcahamam-Çamlıdere Geopark Project in Ankara, Turkey. Geoheritage 4(4):249–261
Keating GN, Valentine GA, Krier DJ, Perry FV (2008) Shallow plumbing systems for small-volume basaltic volcanoes. Bull Volcanol 70(5):563–582
Kereszturi G, Németh K (2012a) Monogenetic basaltic volcanoes: genetic classification, growth, geomorphology and degradation. In: Németh K (ed) Updates in volcanology—new advances in understanding volcanic systems. inTech Open, Rijeka, Croatia, pp 3–88 [http://dx.doi.org/10.5772/51387]
Kereszturi G, Németh K (2012b) Structural and morphometric irregularities of eroded Pliocene scoria cones at the Bakony-Balaton Highland volcanic field, Hungary. Geomorphology 136(1):45–58
Kereszturi G, Németh K, Csillag G, Balogh K, Kovács J (2011) The role of external environmental factors in changing eruption styles of monogenetic volcanoes in a Mio/Pleistocene continental volcanic field in western Hungary. J Volcanol Geoth Res 201(1–4):227–240
Kereszturi G, Geyer A, Marti J, Nemeth K, Javier Doniz-Paez F (2013) Evaluation of morphometry-based dating of monogenetic volcanoes-a case study from Bandas del Sur, Tenerife (Canary Islands). Bull Volcanol 75(7)
Kervyn M, Ernst GGJ, Carracedo JC, Jacobs P (2012) Geomorphometric variability of "monogenetic" volcanic cones: evidence from Mauna Kea, Lanzarote and experimental cones. Geomorphology 136(1):59–75
Kilburn CRJ (1981) Pahoehoe and Aa-Lavas—a discussion and continuation of the model of Peterson and Tilling. J Volcanol Geoth Res 11(2–4):373–382
Koulakov I, El Khrepy S, Al-Arifi N, Sychev I, Kuznetsov P (2014) Evidence of magma activation beneath the Harrat Lunayyir basaltic field (Saudi Arabia) from attenuation tomography. Solid Earth 5 (2):873–882

Le Pennec JL, Temel A, Froger JL, Sen S, Gourgaud A, Bourdier JL (2005) Stratigraphy and age of the Cappadocia ignimbrites, Turkey: reconciling field constraints with paleontologic, radiochronologic, geochemical and paleomagnetic data. J Volcanol Geoth Res 141(1–2):45–64

Lefevre R, Gaudichet A, Leguern F (1991) Hawaiian Lava Fountains—a new source of Glassy Microspherules, various in chemical-composition, for atmospheric aerosols—comparison with Etna. Comptes Rendus De L Academie Des Sciences Serie Ii 313 (1):105–111

Lexa J, Seghedi I, Nemeth K, Szakacs A, Konecny V, Pecskay Z, Fueloep A, Kovacs M (2010) Neogene-Quaternary volcanic forms in the Carpathian-Pannonian region: a review. Cent Eur J Geosci 2 (3):207–275

Lorenz V (2007) Syn- and posteruptive hazards of maar-diatreme volcanoes. J Volcanol Geoth Res 159(1–3):285–312

Major JJ, Pierson TC, Hoblitt RP, Moreno H (2013) Pyroclastic density currents associated with the 2008-2009 eruption of Chaiten Volcano (Chile): forest disturbances, deposits, and dynamics. Andean Geol 40(2):324–358

Mangan MT, Cashman KV (1996) The structure of basaltic scoria and reticulite and inferences for vesiculation, foam formation, and fragmentation in lava fountains. J Volcanol Geoth Res 73:1–18

Manville V, Németh K, Kano K (2009) Source to sink: A review of three decades of progress in the understanding of volcaniclastic processes, deposits, and hazards. Sediment Geol 220:136–161

Martin U, Nemeth K (2006) How Strombolian is a "Strombolian" scoria cone? Some irregularities in scoria cone architecture from the Transmexican Volcanic Belt, near Volcan Ceboruco, (Mexico) and Al Haruj (Libya). J Volcanol Geoth Res 155(1–2):104–118

McKnight SB, Williams SN (1997) Old cinder cone or young composite volcano?: The nature of Cerro Negro, Nicaragua. Geology 25(4):339–342

McPhie J, Doyle M, Allen R (1993) Volcanic Textures. A guide to the interpretation of textures in volcanic rocks. Tasmanian Government Printing Office, Tasmania, p 196

Miallier D, Boivin P, Deniel C, Gourgaud A, Lanos P, Sforna M, Pilleyre T (2010) The ultimate summit eruption of Puy de Dome volcano (Chaine des Puys, French Massif Central) about 10,700 years ago. CR Geosci 342(11):847–854

Moufti MR, Hashad MH (2005) Volcanic hazards assessment of Saudi Arabian Harrats: geochemical and isotopic studies of selected areas of active Makkah-Madinah-Nafud (MMN) volcanic rocks. Final project Report (LGP-5-27) 679 submitted to King Abdulaziz City for Science and Technology, Riyadh, Saudi Arabia, 1–401

Moufti MR, Matsah MI, Soliman MA, Moghazi AM (2011) Arabian plume dynamics beneath Al-Madinah Al-Munawwarah region and its related geohazards. Final project Report (ARP-26–79) submitted to King Abdulaziz City for Science and Technology, Riyadh, Saudi Arabia, pp 1–382

Moufti MR, Moghazi AM, Ali KA (2012) 40Ar/39Ar geochronology of the Neogene-Quaternary Harrat Al-Madinah intercontinental Volcanic Field, Saudi Arabia: implications for duration and migration of volcanic activity. J Asian Earth Sci 62:253–268

Moufti MR, Németh K (2013a) The intra-continental Harrat Al Madinah volcanic field, Western Saudi Arabia: a proposal to establish Harrat Al Madinah as the first volcanic geopark in the Kingdom of Saudi Arabia. Geoheritage 5(3):185–206

Moufti MR, Németh K (2013b) The establishment of a volcanic geopark in the Kingdom of Saudi Arabia: Harrat Al Madinah Volcanic Geopark. In: Jörg Büchner, Vladislav Rapprich, Olaf Tietz (Editors): Basalt 2013—Senckenberg scientific conference [24–28 April 2013, Goerlitz, Germany] Abstracts & Excursion Guides:203–204

Moufti MR, Németh K (2013c) Geosites of the Al Madinah Volcanic Geopark as the first geopark proposed to develop in the Kingdom of Saudi Arabia. In: Abstract volume of the 8th international conference in geomorphology [Paris, France, 27th–31th August 2013] Abstract Number:522

Moufti MR, Moghazi AM, Ali KA (2013a) Ar-40/Ar-39 geochronology of the Neogene-Quaternary Harrat Al-Madinah intercontinental volcanic field, Saudi Arabia: implications for duration and migration of volcanic activity. J Asian Earth Sci 62:253–268

Moufti MR, Németh K, Murcia H, Al-Gorrry SF, Shawali J (2013b) Scientific basis of the geoheritage and geotouristic values of the 641 AD Al Madinah eruption site in the Al Madinah Volcanic Field, Kingdom of Saudi Arabia. Open Geol J [Bentham Science Publisher] 7:31–44

Moufti MR, Németh K, El-Masry N, Qaddah A (2013c) Geoheritage values of one of the largest maar craters in the Arabian Peninsula: the Al Wahbah Crater and other volcanoes (Harrat Kishb, Saudi Arabia). Cent Eur J Geosci 5(2):254–271

Moufti MR, Nemeth K, Murcia H, Lindsay JM (2013d) The 1256 AD Al Madinah historic eruption geosite as the youngest volcanic chain in the Kingdom of Saudi Arabia. Int J Earth Sci 102(4):1069–1070

Moufti MR, Nemeth K, Murcia H, Lindsay JM, El-Masry N (2013e) Geosite of a steep lava spatter cone of the 1256 AD, Al Madinah eruption, Kingdom of Saudi Arabia. Cent Eur J Geosci 5(2):189–195

Moufti MR, Németh K (2014) The White Mountains of Harrat Khaybar, Kingdom of Saudi Arabia. Int J Earth Sci 103(6):1641–1643

Mukhopadhyay B, Mogren S, Mukhopadhyay M, Dasgupta S (2013) Incipient status of dyke intrusion in top crust—evidences from the Al-Ays 2009 earthquake swarm, Harrat Lunayyir, SW Saudi Arabia. Geomatics Nat Hazards Risk 4(1):30–48

Murcia H, Németh K, Lindsay J, Moufti MR, El-Masry N, Cronin SJ, Smith IEM (2013) Complex isolated-to-nested pyroclastic cones and cone rafting in the AD 1256 Al-Madinah fissure eruption in the Kingdom of Saudi Arabia. In: Jörg Büchner, Vladislav Rapprich, Olaf Tietz (Editors): Basalt 2013–Senckenberg scientific conference [24–28 April 2013, Goerlitz, Germany] Abstracts & Excursion Guides:174–175

Murcia H, Németh K, Moufti MR, Lindsay JM, El-Masry N, Cronin SJ, Qaddah A, Smith IEM (2014) Late Holocene lava flow morphotypes of northern Harrat Rahat, Kingdom of Saudi Arabia: implications for the description of continental lava fields. J Asian Earth Sci 84:131–145

Murcia H, Németh K, El-Masry NN, Lindsay JM, Moufti MRH, Wameyo P, Cronin SJ, Smith IEM, Kereszturi G (2015) The Al-Du'aythah volcanic cones, Al-Madinah City: implications for volcanic hazards in northern Harrat Rahat, Kingdom of Saudi Arabia. Bull Volcanol 77(6)

Németh K (2010) Monogenetic volcanic fields: origin, sedimentary record, and relationship with polygenetic volcanism. Geol Soc Am Spec Pap 470:43–66

Németh K, Cronin SJ (2008) Volcanic craters, pit craters and high-level magma-feeding systems of a mafic island-arc volcano: Ambrym, Vanuatu, South Pacific. In: Thomson K, Petford N (eds) Structure and emplacement of high-level magmatic systems. Geological Society, London, pp 85–99

Németh K, Cronin SJ (2011) Drivers of explosivity and elevated hazard in basaltic fissure eruptions: The 1913 eruption of Ambrym Volcano, Vanuatu (SW-Pacific). J Volcanol Geoth Res 201(1–4):194–209

Németh K, Martin U, Haller MJ, Risso C, Massaferro G (2005) Some irregularity in scoria cone degradation in lava spatter-dominated cones. In: VI. International geomorphological world conference. Zaragoza, Spain, p 212

Németh K, Murcia H, Qaddah A, Moufti MRH, El-Masry N, Kereszturi G (2013) Volcano-morphology and eruptiuve processes associated with the historic eruption sites of northern Harrat Rahat, Kingdom of Saudi Arabia. Book of Extended Abstracts—VORISA Scientific Meeting 17–18 November 2013. Faculty of Earth

Sciences, King Abdulaziz University, Geological Hazards Research Unit, Jeddah, pp 29–33
Németh K, Risso C, Nullo F, Kereszturi G (2011) The role of collapsing and cone rafting on eruption style changes and final cone morphology: Los Morados scoria cone, Mendoza, Argentina. Cent Eur J Geosci 3(2):102–118
Németh K, Risso C, Nullo F, Smith IEM, Pecskay Z (2012) Facies architecture of an isolated long-lived, nested polygenetic silicic tuff ring erupted in a braided river system: The Los Loros volcano, Mendoza, Argentina. J Volcanol Geoth Res 239:33–48
Németh K, Suwesi KS, Peregi Z, Gulácsi Z, Ujszászi J (2003) Plio/Pleistocene flood basalt related scoria and spatter cones, rootless lava flows, and pit craters, Al Haruj Al Abyad, Libya. Geolines J Geol Insts AS Czech Repub 15:98–103
Njome MS, Suh CE, Sparks RSJ, Ayonghe SN, Fitton JG (2008) The Mount Cameroon 1959 compound lava flow field: morphology, petrography and geochemistry. Swiss J Geosci 101(1):85–98
Okubo CH, Martel SJ (1998) Pit crater formation on Kilauea volcano, Hawaii. J Volcanol Geoth Res 86(1–4):1–18
Ort MH, Elson MD, Anderson KC, Duffield WA, Samples TL (2008) Variable effects of cinder-cone eruptions on prehistoric agrarian human populations in the American southwest. J Volcanol Geoth Res 176(3):363–376
Pallister JS, McCausland WA, Jonsson S, Lu Z, Zahran HM, El Hadidy S, Aburukbah A, Stewart ICF, Lundgren PR, White RA, Moufti MRH (2010) Broad accommodation of rift-related extension recorded by dyke intrusion in Saudi Arabia. Nat Geosci 3(10):705–712
Parcheta CE, Houghton BF, Swanson DA (2012) Hawaiian fissure fountains 1: decoding deposits-episode 1 of the 1969–1974 Mauna Ulu eruption. Bull Volcanol 74(7):1729–1743
Parfitt EA, Wilson L (1995) Explosive volcanic-eruptions.9. The transition between Hawaiian-style lava fountaining and Strombolian explosive activity. Geophys J Int 121(1):226–232
Parfitt EA, Wilson L (1999) A Plinian treatment of fallout from Hawaiian lava fountains. J Volcanol Geoth Res 88(1–2):67–75
Parfitt EA, Wilson L, Neal CA (1995) Factors influencing the height of Hawaiian lava fountains: implications for the use of fountain height as an indicator of magma gas content. Bull Volcanol 57(6):440–450
Patrick MR, Orr TR (2012) Rootless shield and perched lava pond collapses at KA ≪ lauea Volcano, Hawai'i. Bull Volcanol 74(1): 67–78
Peterson DW, Tilling RI (1980) Transition of basaltic lava from Pahoehoe to Aa, Kilauea Volcano, Hawaii—field observations and key factors. J Volcanol Geoth Res 7(3–4):271–293
Petrovic MD, Vasiljevic DA, Vujicic MD, Hose TA, Markovic SB, Lukic T (2013) Global geopark and candidate—comparative analysis of Papuk Mountain Geopark (Croatia) and Fruska Gora Mountain (Serbia) by using GAM model. Carpathian J Earth Environ Sci 8(1):105–116
Pioli L, Erlund E, Johnson E, Cashman K, Wallace R, Rosi M, Granados HD (2008) Explosive dynamics of violent Strombolian eruptions: the eruption of Paricutin Volcano 1943–1952 (Mexico). Earth Planet Sci Lett 271(1–4):359–368
Pittari A, Cas RAF, Edgar CJ, Nichols HJ, Wolff JA, Marti J (2006) The influence of palaeotopography on facies architecture and pyroclastic flow processes of a lithic-rich ignimbrite in a high gradient setting: The Abrigo Ignimbrite, Tenerife, Canary Islands. J Volcanol Geoth Res 152(3–4):273–315
Polacci M, Papale P (1997) The evolution of lava flows from ephemeral vents at Mount Etna: Insights from vesicle distribution and morphological studies. J Volcanol Geoth Res 76(1–2):1–17
Porter SC (1972) Distribution, morphology, and size frequency of cinder cones on Mauna Kea volcano, Hawaii. Geol Soc Am Bull 83:3613–3626
Powers S (1916) Explosive ejectamenta of Kilauea. Am J Sci 41 (243):227–244
Pulido Fernandez M, Lagar Timon D, Garcia Marin R (2014) Geosites inventory in the Geopark Villuercas-Ibores-Jara (Extremadura, Spain): a proposal for a new classification. Geoheritage 6(1):17–27
Riggs N, Carrasco-Nunez G (2004) Evolution of a complex isolated dome system, Cerro Pizarro, central Mexico. Bull Volcanol 66 (4):322–335
Riggs NR, Duffield WA (2008) Record of complex scoria cone eruptive activity at Red Mountain, Arizona, USA, and implications for monogenetic mafic volcanoes. J Volcanol Geoth Res 178(4):763–776
Riggs NR, Hurlbert JC, Schroeder TJ, Ward SA (1997) The interaction of volcanism and sedimentation in the proximal areas of a mid-tertiary volcanic dome field, central Arizona. USA J Sediment Res 67(1):142–153
Robert B, Harris A, Gurioli L, Medard E, Sehlke A, Whittington A (2014) Textural and rheological evolution of basalt flowing down a lava channel. Bull Volcanol 76(6):1–21
Roche O, de Vries BV, Druitt TH (2001) Sub-surface structures and collapse mechanisms of summit pit craters. J Volcanol Geoth Res 105(1–2):1–18
Roggensack K, Hervig RL, McKnight SB, Williams SN (1997) Explosive basaltic volcanism from Cerro Negro volcano: influence of volatiles on eruptive style. Science 277(5332):1639–1642
Rosseel JB, White JDL, Houghton BF (2006) Complex bombs of phreatomagmatic eruptions: Role of agglomeration and welding in vents of the 1886 Rotomahana eruption, Tarawera, New Zealand. J Geophys Res Solid Earth 111(B12)
Rossi MJ (1997) Morphology of the 1984 open-channel lava flow at Krafla volcano, Northern Iceland. Geomorphology 20(1–2):95–112
Rowland SK, Walker GPL (1987) Toothpaste lava: characteristics and origin of a lava structural type transitional between pahoehoe and aa. Bull Volcanol 49(4):631–641
Rowland SK, Walker GPL (1990) Pahoehoe and aa in Hawaii—volumetric flow-rate controls the lava structure. Bull Volcanol 52 (8):615–628
Rowland SK, Jurado-Chichay Z, Ernst WG, Walker GPL (2009) Pyroclastic deposits and lava flows from the 1759-1774 eruption of El Jorullo, Mexico; aspects of "violent strombolian" activity and comparison with Paricutin. Special Publications of the International Association of Volcanology and Chemistry of the Earth's Interior 2:105–128
Runge MG, Bebbington MS, Cronin SJ, Lindsay JM, Kenedi CL, Moufti MRH (2014) Vents to events: determining an eruption event record from volcanic vent structures for the Harrat Rahat, Saudi Arabia. Bull Volcanol 76(3)
Rymer H, de Vries BV, Stix J, Williams-Jones G (1998) Pit crater structure and processes governing persistent activity at Masaya Volcano, Nicaragua. Bull Volcanol 59(5):345–355
Schwarzkopf LM, Schmincke HU, Cronin SJ (2005) A conceptual model for block-and-ash flow basal avalanche transport and deposition, based on deposit architecture of 1998 and 1994 Merapi flows. J Volcanol Geoth Res 139(1–2):117–134
Seghedi I, Helvaci C, Pecskay Z (2015) Composite volcanoes in the south-eastern part of Izmir-Balikesir Transfer Zone, Western Anatolia, Turkey. J Volcanol Geoth Res 291:72–85
Self S, Keszthelyi L, Thordarson T (1998) The importance of pahoehoe. Annu Rev Earth Planet Sci 26:81–110
Sen E, Aydar E, Gourgaud A, Kurkcuoglu B (2002) Initial explosive phases during extrusion of volcanic lava domes: example from rhyodacitic dome of Dikkartin Dag, Erciyes stratovolcano, Central Anatolia, Turkey. Comptes Rendus Geosci 334(1):27–33
Settle M (1979) Structure and emplacement of cinder cone fields. Am J Sci 279(10):1089–1107

Sheth HC, Ray JS, Kumar A, Bhutani R, Awasthi N (2011) Toothpaste lava from the Barren Island volcano (Andaman Sea). J Volcanol Geoth Res 202(1–2):73–82
Siebe C, Abrams M, Sheridan MF (1993) Major Holocene block-and-ash fan at the western slope of ice-capped Pico-de-Orizaba Volcano, Mexico—implications for future hazards. J Volcanol Geoth Res 59(1–2):1–33
Siebel W, Schmitt AK, Kiemele E, Danisik M, Aydin F (2011) Acigol rhyolite field, central Anatolia (part II): geochemical and isotopic (Sr-Nd-Pb, delta O-18) constraints on volcanism involving two high-silica rhyolite suites. Contrib Miner Petrol 162(6):1233–1247
Simkin T, Siebert L (1994) Volcanoes of the world. Geoscience Press, Tucson, Arizona, p 349
Smith NJ, Kokelaar BP (2013) Proximal record of the 273 ka Poris caldera-forming eruption, Las Canadas, Tenerife. Bull Volcanol 75(11)
Solana M (2012) Development of unconfined historic lava flow fields in Tenerife: implications for the mitigation of risk from a future eruption. Bull Volcanol 74(10):2397–2413
Stovall WK, Houghton BF, Harris AJL, Swanson DA (2009) Features of lava lake filling and draining and their implications for eruption dynamics. Bull Volcanol 71(7):767–780
Stovall WK, Houghton BF, Gonnermann H, Fagents SA, Swanson DA (2011) Eruption dynamics of Hawaiian-style fountains: the case study of episode 1 of the KA << lauea Iki 1959 eruption. Bull Volcanol 73(5):511–529
Suh CE, Stansfield SA, Sparks RSJ, Njome MS, Wantim MN, Ernst GGJ (2011) Morphology and structure of the 1999 lava flows at Mount Cameroon Volcano (West Africa) and their bearing on the emplacement dynamics of volume-limited flows. Geol Mag 148(1): 22–34
Sulpizio R, Capra L, Sarocchi D, Saucedo R, Gavilanes-Ruiz JC, Varley NR (2010) Predicting the block-and-ash flow inundation areas at Volcan de Colima (Colima, Mexico) based on the present day (February 2010) status. J Volcanol Geoth Res 193(1–2):49–66
Sumner JM (1998) Formation of clastogenic lava flows during fissure eruption and scoria cone collapse: the 1986 eruption of Izu-Oshima Volcano, eastern Japan. Bull Volcanol 60:195–212
Sumner JM, Blake S, Matela RJ, Wolff JA (2005) Spatter. J Volcanol Geoth Res 142(1–2):49–65
Suzuki-Kamata K, Kusano T, Yamasaki K (2009) Fractal analysis of the fracture strength of lava dome material based on the grain size distribution of block-and-ash flow deposits at Unzen Volcano. Jpn Sediment Geol 220(3–4):162–168
Umazano AM, Melchor RN, Bedatou E, Bellosi ES, Krause JM (2014) Fluvial response to sudden input of pyroclastic sediments during the 2008–2009 eruption of the Chaiten Volcano (Chile): the role of logjams. J S Am Earth Sci 54:140–157
Valentine GA (2012) Shallow plumbing systems for small-volume basaltic volcanoes, 2: evidence from crustal xenoliths at scoria cones and maars. J Volcanol Geoth Res 223:47–63
Valentine GA, Gregg TKP (2008) Continental basaltic volcanoes—Processes and problems. J Volcanol Geoth Res 177(4):857–873
Valentine GA, Keating GN (2007) Eruptive styles and inferences about plumbing systems at Hidden Cone and Little Black Peak scoria cone volcanoes (Nevada, USA). Bull Volc 70(1):105–113
Valentine GA, Krier DJ, Perry FV, Heiken G (2007) Eruptive and geomorphic processes at the Lathrop Wells scoria cone volcano. J Volcanol Geoth Res 161(1–2):57–80
Valentine GA, Perry FV, WoldeGabriel G (2000) Field characteristics of deposits from spatter-rich pyroclastic density currents at Summer Coon volcano, Colorado. J Volcanol Geoth Res 104:187–199
van Otterloo J, Cas RAF, Sheard MJ (2013) Eruption processes and deposit characteristics at the monogenetic Mt. Gambier Volcanic Complex, SE Australia: implications for alternating magmatic and phreatomagmatic activity. Bulletin of Volcanology 75(8)
Vernet G, Raynal J-P (2008) The Marsat formation and the CF7 tephra, distal markers of violent trachytic eruption of the Chaine des Puys during the Boreal period. Quaternaire 19(2):97–106
Vespermann D, Schmincke H-U (2000) Scoria cones and tuff rings. In: Sigurdsson H, Houghton BF, McNutt SR, Rymer H, Stix J (eds) Encyclopedia of volcanoes. Academic Press, San Diego, pp 683–694
Vujičić M, Vasiljević ĐA, Marković SB, Hose TA, Lukić T, Hadžić O, Janićević S (2011) Preliminary geosite assessment model (GAM) and its application on Fruška Gora Mountain, potential geotourism destination of Serbia. Acta Geographica Slovenica-Geografski Zbornik 51(2):361–377
Wahab AA, Maaty MAA, Stuart FM, Awad H, Kafafy A (2014) The geology and geochronology of Al Wahbah maar crater, Harrat Kishb, Saudi Arabia. Quat Geochronol 21:70–76
Wantim MN, Suh CE, Ernst GGJ, Kervyn M, Jacobs P (2011) Characteristics of the 2000 fissure eruption and lava flow fields at Mount Cameroon volcano, West Africa: a combined field mapping and remote sensing approach. Geol J 46(4):344–363
Wentworth CK, Macdonald GA (1953) Structures and forms of basaltic rocks in Hawaii. US Geol Surv Bull 98–98
White JDL (1989) Basic elements of maar-crater deposits in the Hopi Buttes volcanic field, Northeastern Arizona, USA. J Geol 97: 117–125
White JDL (1990) Depositional architecture of a maar-pitted playa—sedimentation in the Hopi Buttes Volcanic Field, Northeastern Arizona, USA. Sediment Geol 67(1–2):55–84
White JDL (1991) The depositional record of small, monogenetic volcanoes within terrestrical basins. In: Fisher RV, Smith GA (eds) Sedimentation in volcanic settings. Society for Sedimentary Geology, Tulsa (Oklahoma), pp 155–171
White JDL, Bryan SE, Ross PSR, Self S, Thordarson T (2009) Physical volcanology of continental large igneous provinces: update and review. In: Thordarson T, Self S, Larsen G, Rowland SK, Hoskuldsson A (eds) Studies in volcanology: The legacy of George Walker. Geological Society, London, London, UK, pp 291–321
Wolff JA, Sumner JM (2000) Lava fountains and their products. In: Sigurdsson H, Houghton BF, McNutt SR, Rymer H, Stix J (eds) Encyclopedia of volcanoes. Academic Press, San Diego, pp 321–329
Wood CA (1980a) Morphometric analysis of cinder-cone degradation. J Volcanol Geotherm Res 8(2–4):137–160
Wood CA (1980b) Morphometric evolution of cinder cones. J Volcanol Geotherm Res 7:387–413
Woodcock D, Harris A (2006) The dynamics of a channel-fed lava flow on Pico Partido volcano, Lanzarote. Bull Volcanol 69(2):207–215
Xu W, Jonsson S (2014) The 2007–8 volcanic eruption on Jebel at Tair island (Red Sea) observed by satellite radar and optical images. Bull Volcanol 76(2)
Yamashina K, Shimizu H (1999) Crustal deformation in the mid-May 1991 crisis preceding the extrusion of a dacite lava dome at Unzen volcano, Japan. J Volcanol Geoth Res 89(1–4):43–55
Zahran HM, McCausland WA, Pallister JS, Lu Z, El-Hadidy S, Aburukba A, Schawali J, Kadi K, Youssef A, Ewert JW, White RA, Lundgren P, Mufti M, Stewart IC, Anonymous (2009) Stalled eruption or dike intrusion at Harrat Lunayyir, Saudi Arabia? Eos, Transactions, American Geophysical Union 90(52, Suppl):Abstract V13E-2072
Zimmer BW, Riggs NR, Carrasco-Nunez G (2010) Evolution of tuff ring-dome complex: the case study of Cerro Pinto, eastern Trans-Mexican Volcanic Belt. Bull Volc 72(10):1223–1240
Zobin VM, Al-Amri AM, Fnais M (2013) Seismicity associated with active, new-born, and re-awakening basaltic volcanoes: case review and the possible scenarios for the Harraat volcanic provinces, Saudi Arabia. Arab J Geosci 6(2):529–541

4 Volcanic Geoheritage of Other Harrats of Kingdom of Saudi Arabia

4.1 Geoheritage Value of Other Volcanic Fields in Western Saudi Arabia: Overview

Other than the intensive volcanological and geoheritage studies on Harrat Rahat (Moufti and Németh 2013; Moufti et al. 2013b; Murcia et al. 2014, 2015), only preliminary research has been done to document the volcanic geoheritage value of harrats in western Saudi Arabia. Several research paper and conference presentations published between 2012 and 2015 raised awareness of the opportunity to develop the geoheritage, geotourism and geoconservation works in many volcanic regions in the Arabian Peninsula (Moufti et al. 2013a, 2015; Moufti and Nemeth 2014). Preparing the initial volcanic geoheritage catalogue yielded some additional sites to be recognized and recommended for future investigations. The following descriptions result from lengthy and intensive field observations in the various harrats.

Harrat Kishb is a volcanic field in the south of Harrat Rahat (Fig. 4.1) which provides new insight into the evolution of a volcanic field that shows more similarity to normal intracontinental volcanism with bimodal (basalt and phonolite) alkaline volcanism (Camp et al. 1992; Coleman and Gregory 1983; Connor and Conway 2000). The first visits to Harrat Kishb were not only to understand the geological context of the region but also document and catalogue the volcanic geoheritage values of the region. Harrat Kishb is the home of two particular volcanic sites regularly visited by tourists in the past and initial concern about the preservation of sites with high geological heritage values (Moufti et al. 2013a) was raised by the Saudi Authority for Tourism and Antiquity. These two sites are considered to be the two most prominent eruption sites of the field; (1) the Al Wahbah maar crater (Fig. 4.2) and the (2) Aslaj volcanic complex (Fig. 4.3).

The SE edge of Harrat Kishb also has potential for geoheritage development (Moufti et al. 2013a); in that area phonolitic volcanoes (lava domes, dome coulees, phonolitic explosion craters) and large complex volcanic explosion craters are surrounded by tuff rings and nested by lava spatter and scoria cones and associated intra-crater to breached lava flows (Fig. 4.4). These regions are covered by extensive ash plains, which provides dramatic volcanic landscapes and graphic views to help to understand the consequences of major explosive volcanic eruptions. In this regard this region is considered an ideal place to develop volcanic hazard educational sites.

Some preliminary field observations were also taken from Harrat Hutaymah in the northern side of the Arabian Peninsula (Fig. 4.1), which is a prominent and less known volcanic field north of Harrat Rahat (Moufti et al. 2015). Harrat Hutaymah is located near to the city of Hail in Northern Saudi Arabia and has appropriate infrastructure for accessing excellent volcanic sites. In Harrat Hutaymah visitors can see some of the best sites in western Saudi Arabia for illustrating the eruption mechanism of explosion craters dominated by phreatomagmatic explosive eruptions (Moufti et al. 2015). Harrat Hutaymah is also an excellent site to contribute to our understanding of how maar craters form and how and to what extent we can relate maar volcanism and pre-maar scoria and lava spatter-forming eruptions which are commonly fissure aligned (Fig. 4.5).

The volcanic field is ideally located in the proximity of Hail city, allowing for day tours from the city. In addition to the volcanic past, rich archaeological remains document the development of irrigation in the region over the past several thousands of years history. Thus this location is a unique place to combine geological and cultural heritage studies and to develop geotouristic and geoeducational sites (Moufti et al. 2015).

Harrat Khaybar, just north of the city of Al Madinah (Fig. 4.1), has been very remote, but new road construction in the region in combination with the proximity to other cultural heritage sites, including the Mada'in Saleh UNESCO World heritage site, provide the opportunity for this field to

© Springer International Publishing Switzerland 2016

M.R. Moufti and K. Németh, *Geoheritage of Volcanic Harrats in Saudi Arabia*,
Geoheritage, Geoparks and Geotourism, DOI 10.1007/978-3-319-33015-0_4

Fig. 4.1 Harrats other than Harrat Rahat with volcanic geoheritage values on google earth image. Individual harrats described in this chapter are marked on satellite image such as Harrat Kishb, Harrat Khaybar, Harrat Hutaymah and Harrat al Birk

Fig. 4.2 Al Wahbah maar crater overview from the western scoria cone half sectioned by the maar-forming eruption [22° 54′ 37.69″N; 41° 8′ 2.98″E]

be one of the flagship regions to form a volcanic geopark in a unique volcanic landscape. The advance of a new road from the township of Khaybar to the east allows access to the deep interiors of the field across a relatively large area (Fig. 4.6). The purpose of the preliminary field campaign here was to locate volcanic geoheritage sites, and identify major geological features that make this harrat unique. Harrat Khaybar is one of the largest harrats in western Saudi Arabia and hosts nearly as many eruptive centers as Harrat Rahat, with a substantially larger number of silicic centers that are well-preserved (Camp et al. 1991).

The Harrat Khaybar is inferred to have had volcanic activity during historic time (Camp et al. 1991). The largest volcano preserved in Harrat Khaybar (Jabal Qidr) is one of the youngest (Fig. 4.7) and produced an extensive ash plain (Camp et al. 1991) suspected to have resulted from a

Fig. 4.3 Aslaj is a complex volcano surrounded by a volcanic ash plain in the norhtern margin of the Harrat Kishb [23° 14′ 17.19″N; 41° 16′ 6.64″E]

Fig. 4.4 Phonolithic volcanoes of Harrat Kishb looking toward the west/north-west from the southern margin of the volcanic field [22° 48′ 28.14″ N; 41° 20′ 53.92″E]

Fig. 4.5 Harrat Hutaymah overview and landscape [26° 58′ 57.50″N; 42°14′ 45.61″E]

Fig. 4.6 Typical dirt road in the interior of Harrat Khaybar with having Jebel Bayda in the background [25° 39′ 51.45″N; 39° 57′ 26.51″E]

Fig. 4.7 Jebel Quidr viewed from the east [25° 43′ 12.19″N; 39° 56′ 37.64″E]

Fig. 4.8 Harrat al Birk landscape looking toward south along the Red sea in the outskirt of Al Birk township [18° 12′ 48.73″N; 41° 32′ 17.94″E]

sustained eruption (active longer than few months) that likely reached sub-Plinian scales and accompanied the growth of a stratovolcano (Martin and Németh 2006; Valentine and Gregg 2008). Due to the young suspected age of Jabal Quidr the need to collect information to evaluate the volcanic hazard in the near future in the vicinity of the city of Al Madinah is justified. The presence of the intact volcanic landforms in the harrat can be used as geoeducation sites to demonstrate the causes and consequences of such eruptions in the future.

Harrat Al Birk is located in the southwest of Saudi Arabia, along the Tihamat Asir region near the Red Sea coast (Fig. 4.1). The Harrat Al Birk contains some spectacular volcanic landscapes that are particularly beautiful due to the proximity of the volcanoes to the Red Sea (Fig. 4.8). The volcanic geoheritage value of Harrat Al Birk is high and it is also justified by the fact that some historic eruptions are known from this region and the current dramatic economic development in the region accompanied with fast population growth would make perfect sense to continue a volcanic hazard study and associated geoheritage and geoeducational work to be conducted in the region in the near future.

4.2 Harrat Kishb

4.2.1 Al Wahbah Maar Crater Geotope [22° 54′ 2.11″N; 41° 8′ 23.36″E]

Al Wahbah Crater is part of the Harrat Kishb (Fig. 4.9), a bimodal (alkaline olivine basalt–phonolite) monogenetic volcanic field with extensive lava fields (harrat) formed in the last 2 millions of years (Abdel Wahab et al. 2014; Camp et al. 1992; Coleman and Gregory 1983; Grainger 1996; Grainger and Hanif 1989). Al Wahbah Crater is one of the largest and deepest Quaternary maar craters in the Arabian Peninsula (Fig. 4.9). It is NW-SE-elongated, ~2.3 km wide, ~250 m deep and surrounded by an irregular near-perpendicular crater wall cut deeply into the Proterozoic diorite basement (Fig. 4.10). The main feature of the Al Wahbah crater beside its impressive deep crater is the half section of a scoria cone dissected by the crater floor subsidence of the maar, exposing the entire inner part of the scoria cone (Fig. 4.10). The age of Al Wahbah is poorly constrained, however a recent Ar/Ar study on a dolerite plug filling the half sectioned scoria cone confirmed a minimum age of Al Wahbah to be 1.147 ± 0.004 Ma, younger than the uppermost pre-maar lava flow emplacement period (1.261 ± 0.021 Ma–1.178 ± 0.007 Ma) (Abdel Wahab et al. 2014). While the age difference between the pre-maar lava fields and the maar appears to be too long (at least 20 ka) to view the maar and its underlying volcanic succession as a result of the same volcanic episodes, there is geological evidence indicating that at least the uppermost pre-scoria cone/pre-maar lava flow is in close genetic relationship with the scoria cone sitting on it.

Maar craters are relatively rare (or rarely preserved due to quick eolian crater fill formation) features of the arid Arabian landscape, and with their "hole-in-the-ground" morphology (Heiken 1971) they differ strikingly from other volcanic landforms in the region; this adds significant and unique landscape value to the volcanic fields of Arabia (Camp et al. 1991, 1992). Al Wahbah is not only unique by its landscape, but also its volcanic geology is special. Pre-maar scoria cone-building rock units and phreatomagmatic successions associated directly with the formation and growth of the maar crater are perfectly preserved. Similar examples are rare globally, particularly ones that are perfectly exposed and/or accessible (Keating et al. 2008; Valentine and Cortes 2013; Valentine and Gregg 2008).

The Proterozoic diorite basement is covered by at least two basaltic lava sheets exposed everywhere in the crater wall (Fig. 4.11). They are dark, aphanitic basanitic rocks with olivine and pyroxene phenocrysts. The lower lava flow is hard to access, and it can be examined only from distant photographs or an access path from the northern edge of the crater rim which leads to the crater floor and an old and abandoned terraced garden in the NE crater wall. The lowermost lava flow unit is columnar jointed thick (10–15 m) lava that thins toward the south. Its base cannot be seen due to cover from debris while its upper boundary to the upper lava flow is separated by lava top, lava foot breccias, and some mixed dusty sedimentary layers that indicate a potential time break between the emplacement of the two flow units. In the upper flow unit (about 5–10 m thick) the lava foot breccias occasionally developed over fairly thick intermediate volcaniclastic bedded sediments in which some pillowed lava lobes can be recognized indicating that the topmost lava flow may have been emplaced in a terrain contained some pond with accumulating volcaniclastic dust from various distal sources. The topmost part of the upper lava flow unit (Fig. 4.12) appears to be multiple lava flow units, including an upper lava flow unit that is clearly thinning from the pre-crater scoria cone and which is the basal part of the scoria cone cut subsequently by the maar crater.

This topmost basal lava flow seems to evolve gradually to a succession that is dominated by lava spatter horizons, clastogenic lava flows and agglutinated scoria beds as the base of the pre-maar scoria cone. The scoria cone edifice up section shows more evidence of a loose scoria ash and lapilli hosted occasional spindle bomb-bearing pyroclastic sequence that is gradually transformed in distal areas to a grain flow-dominated reworked talus that results from a growing scoria cone as a response to the enlarged edifice when the freshly erupted pyroclasts roll down to the foothill of the cone for their resting point. The scoria cone succession is capped by a fine lapilli tuff and tuff succession that is rich in accidental lithic fragments from country rocks and chilled blocky juvenile pyroclasts indicating that they formed due to phreatomagmatic fragmentation and underground explosion triggered country rock fragment excavation (Fig. 4.13a, b). The uppermost tuff ring-forming pyroclastic succession is up to 30 m thick in the NNE and NNW side of the crater located just in the side of the half-sectioned pre-maar scoria cone.

The tuff ring succession consists of a coarse grained basal succession that is rich in accidental lithic fragments and large juvenile pyroclasts with impact sags and it is stratified (Fig. 4.13b). In the middle of the section in the NNE crater wall a thick massive topography filling lapilli tuff forms a prominent unit that is interpreted to be a result of a high particle concentration "moist" pyroclastic density current, e.g. similar to pyroclastic flows (Fig. 4.14). The uppermost pyroclastic succession has more prominent dunes and cross stratification (Fig. 4.15) and abundant evidence of plastering, ballistic bomb emplacement, accretionary lapilli beds and a typical coarse—fine—coarse repetition of tuff lapilli

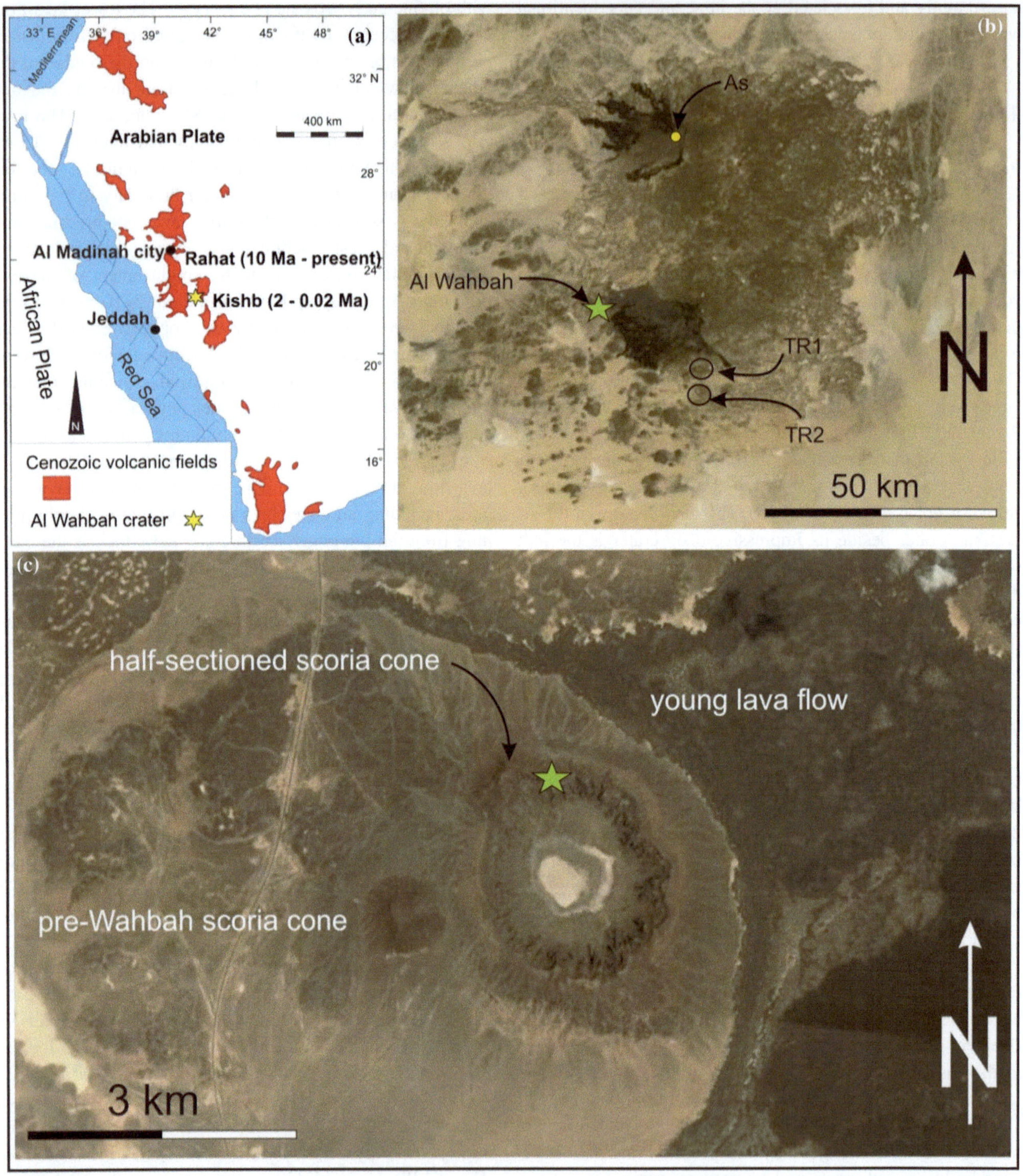

Fig. 4.9 Overview map of the Harrat Kishb from google earth image **b** marking Al Wahbah Crater, Aslaj volcano (*As*) and two other large tuff ring (TR1, TR2) in the southern margin of the volcanic field. On **c** Al Wahbah crater is seen with a *green star* marks the main section of the tuff ring succession

tuff units indicating a stage in the eruption when the erupting conduit was fairly stable and open (e.g. the vent clearing was completed and the maar crater was probably already formed). Interestingly the tuff ring succession is different in the NNW side where the upper succession of dune bedded tuff is more prominent and thicker, and fine ash dunes can be traced over a km from the crater rim. The tuff ring succession thins toward the south, however in the most southernmost

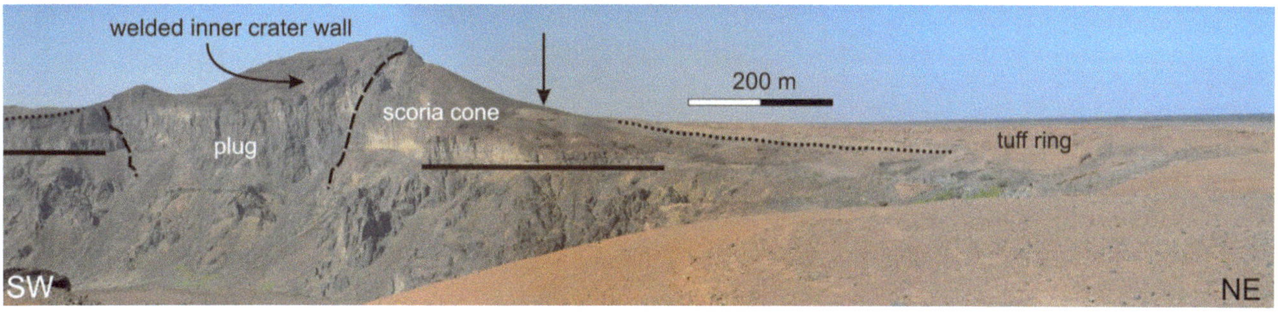

Fig. 4.10 Half-section of a pre-maar scoria cone [22° 54′ 31.84″N; 41° 7′ 59.93″E] in the NW crater wall of Al Wahbah. The pre-maar scoria cone sits on at least two major lava flow units

Fig. 4.11 Topmost pre-maar lava flow unit with multiple individual flow units and undulating upper surface [22° 54′ 12.93″N; 41° 9′ 2.35″E]. Note the *reddish colour* of the capping tuff ring deposits that form about a 30 m thick pile in the eastern edge of the maar crater

Fig. 4.12 Occasionally the *uppermost* pre-maar lava flow shows apparent intrusive contact with the tuff ring deposits [22° 54′ 39.44″N; 41° 8′ 36.90″E] suggesting that the flow might have been still moving in the time the tuff ring-forming deposits accumulated (*bottom image*)

part of the maar crater rim it is again became slightly thicker providing an impression that the tuff ring around the gradually forming maar crater must have been irregular in thickness and potentially indicates that the source of the tuff ring might have been along a fissure allowing to form complex tuff ring around the finally formed maar basin.

There is no clear evidence to support significant time break between the pre-maar cone-building eruptions and the tuff ring-forming base surge-dominated successions, suggesting that pre-maar cone (and lava flow(s), at least their upper part) and the tuff ring formation might be in a time-continuum and part of the same eruptive episode. In this respect Al Wahbah's eruption can be seen as unique and differs from other maar formation implying initial magma-water interaction-driven shallow sub-surface explosions followed by gradual explosion locus down-migration as a result of gradual exhaustion of ground-water sources (Lorenz 1974, 1986). Al Wahbah seems to have followed an opposite eruption evolution starting with an initial lava shield and cone-building phase that have been intervened by

Fig. 4.13 Images from the tuff ring deposits exposed in the N–NE side of the Al Wahbah maar [22° 54′ 41.88″N 41° 8′ 17.33″E]. **a** The tuff ring deposits developed over the pre-maar scoria cone succession covering it partially. **b** Well-bedded, cross-bedded successions of pyroclastic density current deposit dominated succession forms the majority of the tuff ring succession. Large cross-stratification and dune-bedded nature with large accidental lithic clasts derived from the underlying Proterozoic basement are prominent characteristic features of the tuff ring succession

phreatomagmatic explosive eruptions subsequently that have culminated in a maar collapse and tuff ring formation. This eruption scenario can be best explained in similar way how it was proposed for another spectacular maar, the Crater Elegante in Sonora, Mexico (Gutmann 1976, 2002). The drop of the magma discharge rate is inferred to cause magma withdrawal below the regional ground-water table allowing direct entry of groundwater to the hot interior of the shallow plumbing system of the volcanic complex and triggering phreatomagmatic explosions that formed numerous base surges built the tephra ring (Németh et al. 2013). Country rock excavation and the mechanical destabilisation of the basement and pre-maar volcanic edifices eventually led to subsequent crater floor subsidence.

The flat-floored crater hosts an ephermal lake today that is filled with shallow saline water after occasional heavy rain fall (Fig. 4.16) that can diminish quickly leaving behind saltpans and aeolian silt bars. The large crater also acts as "humidity trap" providing refuge for living creatures from the heat, making Al Wahbah a host of a unique ecosystem (Fig. 4.17).

The crater rim has some advanced erosional features, where ~200 metres retreat of the original tephra ring is apparent due to the erosional strip off the tephra beds from the underlying pre-maar lava flows. To constrain the above proposed model some independent data would be essential to constrain the age relationship between pre-maar lava flow units, the half sectioned scoria cones and the maar-related tuff ring. The recently published Ar–Ar age survey has

Fig. 4.14 Prominent pyroclastic flow bed in the middle of the Al Wahbah tuff ring [22° 54′ 45.51″N; 41° 8′ 25.79″E]

suggested that the time difference between the onset of maar-forming eruption (based on the dolerite plug intruded into the half sectioned scoria cone) and the age of the pre-maar lavas could be in the range of 20 Ka, and thus the two events are separate and there is no link between the maar-forming eruption and the pre-maar scoria cone activity and lava flow effusion (Abdel Wahab et al. 2014). If this is the case the eruption that created the Al Wahbah maar occurred in a place where pre-existing scoria and spatter cones formed just few thousands to tens of thousands of years ago. The violent explosive eruption cut into the pre-existing scoria and spatter cone field exposing in its crater wall those half-sectioned older volcanoes. Interestingly similar situations have recently been documented from the Harrat Hutaymah, and therefore this eruption history can equally be valid (Moufti et al. 2015).

In conclusion Al Wahbah is a maar with a peculiar geological history and a complex volcanic stratigraphy. Its volcanic rock units' stratigraphic position is under intensive research to determine whether the pre-maar scoria cone and topmost pre-maar lava flows erupted immediately prior the maar formation, or if they represent completely different volcanism pre-dating the maar forming eruption. Al Wahbah maar can be seen as a complex geotope. It demonstrates a complex volcanic geology history through well-preserved geosites along its crater wall and crater rim many of them providing spectacular views. The fact that this location merits active research in cutting edge research fields warrants general scientific interest in the site by the volcanology community.

4.2.2 Other Silicic Explosion Craters in Harrat Kishb

As an initial exploration to see the geoheritage diversity of volcanic products of Harrat Kishb, field visits were arranged to other volcanic sites Just SE of Al Wahbah are volcanoes here referred to as TR1 (the northern volcano) and TR2 (the southern volcano) (Fig. 4.9). TR1 is a complex volcano in which a double small scoria cone was visible in its main crater (Fig. 4.18a). TR2 is a broad flat-floored crater sitting on a broad relatively gentle sloped volcanic edifice (Fig. 4.18b). These two sites are about 5 km apart.

TR1 is a positive volcanic landform standing about 70 m above the basement, which is a typical alluvial plain with

Fig. 4.15 *Upper* dune-bedded surge beds [22° 54′ 46.49″N; 41° 8′ 15.04″E]

shallow dry valleys and windblown dust. Its volcanic entity and the associated volcanic geosites made this location a stand-alone volcanic geotope that can offer numerous geosites to understand the formation of a broad flat floored crater and eruption of post-crater formation volcanic products. The volcano is slightly asymmetric and has a peak on its western edge, while in the east its crater rim is open and initiated an extensive lava flow that can be traced over several kilometres toward the east. The lava outflow point is above the basement suggesting that the entire TR1 edifice is sitting on a lava shield and that the eruption and formation of this volcano might have started with effusive activity.

The main edifice building succession of TR1 is exposed in several dry gullies in the southern flank of the volcano and exposes chaotic tuff breccias and lapilli tuff that are rich in silicic pyroclast fragments in an ash matrix. It is inferred to be a result of eruption-fed pyroclastic density currents interbedded with some syn-eruptive reworked volcaniclastic deposits. The upper section of the TR1 edifice is more prominent and consists of finer grained, dune-bedded, cross-bedded tuff and lapilli tuff with occasional accretionary lapilli. The crater of the TR1 is flat-floored filled with aeolian dust and is an oval shape with distinct geometry suggesting that it might have been evolved in multiple events and potentially aligned along a fissure along a feeder dyke length. The central to eastern sector of the crater is filled with intra-crater cones (Fig. 4.19). Some lava spatter bench can be traced in the northern central inner wall of the inner crater wall. This lava spatter is clearly connected to a large intra-crater spatter rampart forming an edge in the southern inner crater wall from the base of the crater to its top. This suggests that some larger lava spatter cone must have been existed in the double crater that was either destructed by some explosions and/or were originally largely asymmetric and only provided some lava spatter to plaster against crater walls indicating changes of lava spatter activity during the eruption. In the centre part of the crater closer to the southern crater wall two circular and well-preserved lava spatter/scoria cones stand in the flat crater floor. They are dominated by agglutinated lava spatter, fusiform bombs and minor scoriaceous ash and lapilli. Their intact morphology suggests that they represent the final eruption episode of the volcano growth.

Logistically the TR1 can be accessed through dry valleys (Fig. 4.20) from Al Wahbah; this takes about an hour as a 4WD trip. Potential hazards are the sand dunes and dust

Fig. 4.16 Ephermal lake in Al Wahbah maar crater [22°54′ 2.12″N; 41° 8′ 20.96″E]

Fig. 4.17 Vegetation in Al Wahbah crater after a rainy day [22° 54′ 9.86″N; 41° 9′ 3.10″E]

storms. The majority of the main sections can be accessed from the southern foothills. From the east through the lava outflow there is a rough 4WD track which allows driving into the crater.

TR2 is a broad gentle sloping volcanic edifice which stands about 100 metres above the basement of alluvial plains (Fig. 4.18b). The crater rim is not breached and therefore there is no access to the crater by 4WD. The volcanic edifice

Fig. 4.18 TR1 tuff ring [22° 48′ 15.58″N; 41° 21′ 12.80″E] in the southeastern side of Harrat Kishb is a positive landform with an elongated crater that is filled with small lava spatter and scoria cones (**a**). TR2 [22° 45′ 45.86″N; 41° 21′ 24.08″E] is also a positive volcanic landform with intact crater rim with no intra-crater edifices (**b**)

of TR2 is surrounded by a broad ash plain that consists of light coloured pumiceous ash. In the northern side of the volcanic edifice shallow gullies have developed as a deep and steep walled gully network near the top of the tuff ring. In these valleys dune-bedded lapilli tuff and tuff are exposed that are interpreted to be deposits of pyroclastic density currents initiated radially from the TR2 volcano.

In the northern flank of TR2 an access road approaches the crater rim, allowing the crater rim to be accessed on foot. The crater is filled with aeolian deposits and a few humps in the crater floor indicate buried intra-crater lava spatter cones or lava buds. The majority of the pyroclastic succession along the volcanic edifice seems fairly uniform and suggests that the formation of this volcano might have been a relatively simple and not to long process that is different from the TR1.

The TR1 and TR2 volcanoes are fairly different from Al Wahbah, as they are more like tuff rings without a characteristic deep crater and form a positive volcanic landform with broad but relatively shallow craters (in comparison with Al Wahbah). Each location can be defined as an individual volcanic geotope that can offer useful comparisons with other tuff rings (Kereszturi and Németh 2012). The unusual aspect of these sites is that they are more silicic in composition (phonolitic) than the majority of the eruptive products known from the field and their eruption history may reflect long lived eruptions fed by silicic magma (Camp et al. 1992). There is some evidence for phreatomagmatism in both volcanoes, but their formations were more dominated by some magmatic explosive processes; e.g. their explosive eruptions were probably triggered by phreatomagmatism but sustained by magmatic volatiles due to their more silicic composition. This question needs to be resolved in the future and it could give a substantial research direction for future work.

4.2.3 Aslaj Volcanic Complex Geotope [23° 14′ 26.06″N; 41° 16′ 1.22″E]

Aslaj is located in the northern part of the Harrat Kishb (Fig. 4.9), slightly off from the main volcanic chain that consists of 7 large cones (Fig. 4.21). Aslaj is associated with an extensive lava field that fed pahoehoe to transitional pahoehoe lava flows toward the west (Guilbaud et al. 2007; Peterson and Tilling 1980; Rossi 1997; Rowland and Walker 1990; Sato 1995; Self et al. 1998). The lava field is partially covered by an ash plain that is wind erosion modified. The Aslaj cone is a complex volcanic cone with a network of pit craters, and a large central crater. The main cone of Aslaj is a large steep sided relatively well-preserved volcanic edifice that stands about 100 m above the base (Fig. 4.22a, b). It sits on a gently sloping "shield-like" edifice that consists of mantle nodule bearing multiple lava lobes and sheets. The main cone has a fairly deep crater that is unusually deep in comparison to its base diameter (Fig. 4.23a, b).

Fig. 4.19 Crater interior of TR1: In *upper* image, note the open crater with a lava initiation point. In the NW side of the crater is flat and rimmed by relatively low tuff ring (*middle*). The intra-crater cone complex stands in the middle of the elongated crater floor (*bottom*) [22° 48′ 13.91″N; 41° 21′ 14.58″E]

The steep sloped pyroclastic beds forming the volcanic edifice are dominated by juvenile pyroclasts of scoria of mostly ash to fine lapilli grain size. The individual beds are a few dm thickness and form fairly uniform facies architecture from base to top. The main cone building succession is covered by a relatively thin but accidental lithic rich pyroclastic unit that is dune-bedded, cross-bedded and rich in impact sags commonly cause by large (dm-scale) mantle nodules (Fig. 4.24). These beds are particularly rich in mantle nodules; however, the main cone building pyroclastic beds also contain mantle-derived xenoliths. In addition, these capping pyroclastic beds are also rich in various deep-crustal fragments and accidental lithics. Fine-grained beds exhibit some accretionary lapilli and prominent plastering effect against obstacles.

The best example of this process is the steeply inclined thin dune bedded pyroclastic unit plastered against the main cone inner crater wall. The thickness of this succession varies greatly along the main cone and is seemingly thicker in the south where the main edifice consists of the lower crater rim that is obscured. The southern part of the volcanic complex is complicated by local pit craters, lava spatter

Fig. 4.20 TR2 tuff ring [22° 45v49.53″N; 41° 21′ 22.72″E] in the southeastrn margin of the Harrat Kishb is in the distance in the panoramic view taken from Tr1. Note the extensive ash plain in the inter-cone areas. In the *right hand side* of the *panoramic view*, phonolitic lava flows and domes form the landscape. *View* is toward the SW

vents and local initiation points of lava flows that erupted toward the south and reach several kilometres downhill from their source. This lava flows are thick and consist of lava tubes and pahoehoe surface textures that quickly transform to more aa-like lava flow surface textures. The base of the lava flows are commonly choked with large (dm-scale) mantle nodules. From the top of Aslaj it is evident that a broad (but wind modified) grey ash plain associated with the volcano is cantered in the slightly eastern edge of the visible ash plain.

Aslaj is potentially an interesting volcano to develop as a geotouristic destination with high volcanic geological value due to its complex volcanic evolution. On the available field data, we can infer that the initial cone growth was followed and/or been penecontemporaneous with the lava shield building phase that gradually caused some sort of cone modification over time. The cone growth must have been not only the result of relatively low energy Strombolian style periodic explosive eruptions, but some higher energy explosive events must have taken place also to generate an unusually deep crater in relative to the cone base. Accompanied with the cone growth lava flows were initiated mostly from the southern side of the cone complex that partially and gradually rafted the growing edifice. This process likely induced local pit crater formation (Okubo and Martel 1998; Rymer et al. 1998) that led to the formation of a complex crater area in the south. This rough terrain has been covered by a pyroclastic deposit with non-uniform thickness that is rich in large mantle nodules. The cross bedding and large dunes, plastering effect, and presence of accretionary lapilli together as characteristic features inferred from many places (Moore et al. 1966; Németh 2010; Vespermann et al. 2000) suggest that a paroxysmal phreatomagmatic explosive eruption must have occurred in the final stage of the formation of Aslaj that generated blast-like base surge-dominated eruptions that deposited a typical pyroclastic density current deposit over the rugged topography.

In addition, this eruption episode likely produced highly fragmented ash (phreatomagmatic) that is the likely deposit of the upper part of the extensive ash plain. While this working model needs to be refined and supported or modified in future field works it can be stated that Aslaj could serve as a very valuable geotouristic destination especially in a form of adventure tourism. In addition, the abundant

Fig. 4.21 Chain of scoria cones in the northern part of Harrat Kishb from about 10,000 m height. Aslaj is a tuff cone that is located in the top left side of the 7 cones chain. Note the phonolitic lava flows and dome in the *right bottom side* of the image [23° 11′ 16.16″N; 41° 22′ 48.28″E]

mantle nodules make this place also a perfect site to develop a complex geoeducational program on understanding alkaline basaltic magma generation, rise and the various style of explosive fragmentation. Aslaj is a very significant site where complex volcanological research could bring significant and fundamental results that could feed into a well-designed geoeducational and geotorustic program. Aslaj in this regard has a high volcanic geoheritage value as a complex volcanic geotope in spite its remoteness.

4.3 Harrat Hutaymah

4.3.1 Overview

Harrat Hutayma consists of small-volume volcanoes typical for an alkali basaltic volcanic field (Fig. 4.25). The field has been mapped in the 80 s to 90 s (Pallister 1985; Thornber 1990; Thornber and Anonymous 1988; Thornber and Pallister 1985) and has an abundance of mantle and crustal-derived nodules primarily recovered from various tuff ring and maar deposits (Duncan et al. 2016; Konrad et al. 2016; Thornber 1992, 1993, 1994; Thornber 1988, 1991; Thornber and Pallister 1985). Relatively little information is available in major international journals about the significance of these nodules (Blusztajn et al. 1995; Gondal et al. 2009; McGuire 1988). Volcanological study that characterizes the style of volcanism, its depositional environment and associated volcanic landscape evolution has not been published from this region other than basic information collected for the preparation of geological maps (Pallister 1985). This lack of information as well as the apparent difference of volcanic landforms visible from satellite imagery in comparison to the Rahat Volcanic Field provided a driving force to arrange pilot research to evaluate the Harrat Hutaymah as a potential future research area and initially catalogue its potential geoheritage value and sites (Moufti et al. 2015).

Harrat Hutaymah contains numerous tuff rings and maars that were reported in the first mapping of the region, which identified these volcanic landforms as the main volcanic features that makes Harrat Hutaymah different from other harrats. In this respect Harrat Hutaymah has shown some common volcanic features with Harrat Kishb which made this field also a good target area to explore the variety of

Fig. 4.22 Aslaj tuff cone complex from the SE (*top*) [23° 14′ 16.78″N; 41° 16′ 24.91″E] and S (*bottom*) exhibit a complex volcanic edifice architecture [23° 13′ 40.65″N; 41° 15′ 47.60″E] indicating long-lived eruptions of various eruption types

Fig. 4.23 Complex crater zone of Aslaj volcano with a deep crater of the main cone [23° 14′ 18.28″N; 41° 16′ 6.98″E]

volcanic landforms and associated volcanic processes that created them. These fields demonstrate the diversity of volcanic fields across the Arabian Peninsula.

Harrat Hutaymah covers an area of about 900 km^2 and is considered to be one of the at least 13 distinct flood lava fields (harrats) of the Arabian Peninsula (Pallister 1985). While lava flows are volumetrically important contributors to the total volume of the volcanic eruptive products of Harrat Hutaymah, the field is differ from Harrat Rahat in that areas covered by lava fields strongly influence access to the region. At Harrat Hutaymah, lava flows are more distinct, commonly follow longitudinal networks of valleys, and their apparent thickness is less than the thick multiple lava flow fields in Harrat Rahat. Eruptive centers are commonly aligned at Harrat Hutaymah and scoria cones form long (over tens of kilometres) N-S oriented chain of cones (Fig. 4.25). In this respect Harrat Hutaymah shows great similarity to Harrat Kishb, having a main aligned chain of cones in its central axis. Volcanic cones also commonly form multiple nested cones and ellipsoid (in map view)

Fig. 4.24 Large lower crustal xenolith from the Aslaj volcano [23° 14′ 13.20″N; 41° 16′ 3.47″E]

Fig. 4.25 Overview GoogleEarth map of the Hutaymah Volcanic Field. The black circles mark volcanic craters with high volcanic geoheritage values. *White circle marks* an additional crater that could be used as a site to demonstrate crater filling processes and its importance to understand volcanic landscape evolution

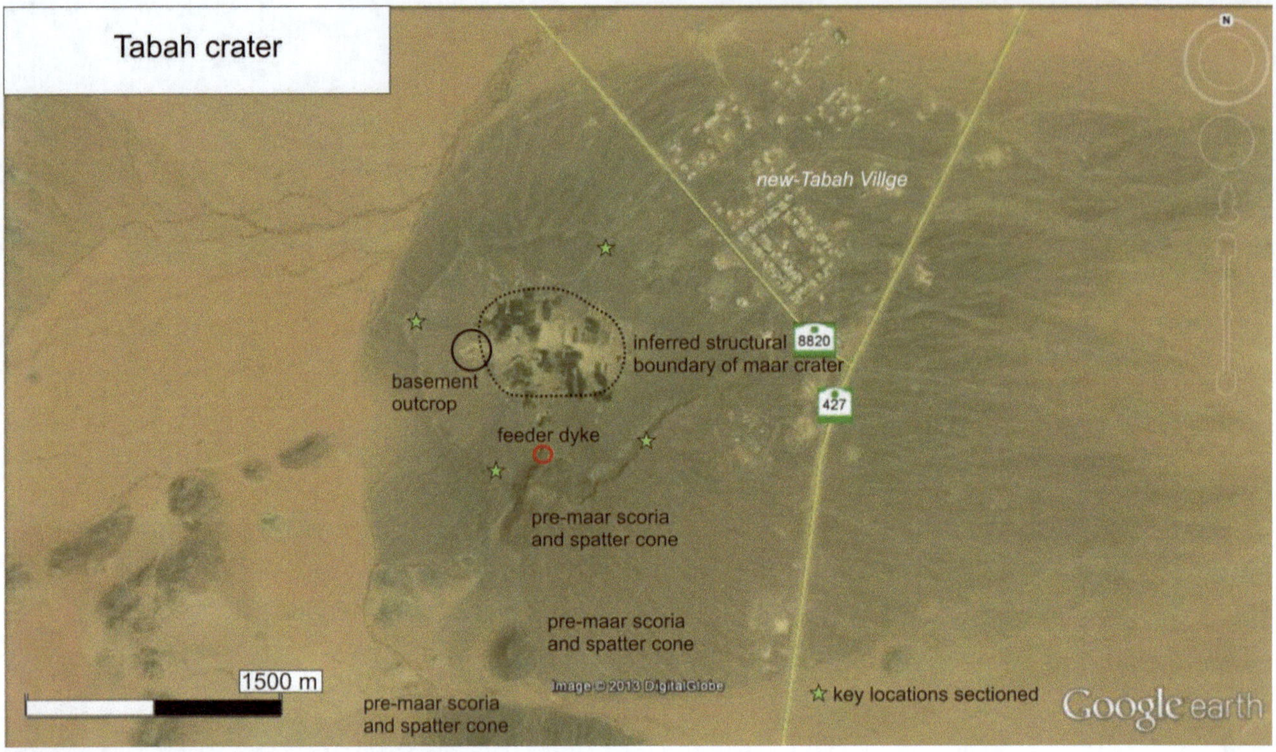

Fig. 4.26 Google earth map of the Tabah crater [27° 1′ 36.34″N; 42° 10′ 6.48″E]. Sections recorded are marked with *green stars*. Note the exposed basement rock in the western side of the present day volcanic depression

shape chain of craters indicating lava curtain-style eruptions along fissures (Pallister 1985). The lava flows are dominantly alkali basaltic flows (Pallister 1985). The age of the flows is poorly constrained having listed Quaternary and probably in the range of 2–1 millions of years age. K-Ar dating from a basal lava flow cut by the Harrat Hutaymah maar yielded an age of 1.8 ± 0.5 Ma (Pallister 1985). Recent age determinations on 14 lava flows by the Ar-40-Ar-39 laser step heating method provided ages between 260–850 Ka, all younger than the previously defined 0.1 to 2.7 Ma age range for the volcanic activity of Harrat Hutaymah (Duncan et al. 2016).

The field contains at least 57 relatively small scoria and lava spatter cones (Pallister 1985) which are smaller than those in Harrat Rahat. At least 22 tuff rings and maars were identified in this field (Pallister 1985; Thornber 1990); however, very little detail is given about their volcanic facies architecture and their inferred eruption mechanism. The field is covered by aeolian deposits that make it difficult to identify volcanic landforms especially in the southern margin of the fields where volcanic craters are entirely filled with such deposits and the crater rims are commonly completely eroded (Thornber 1990). In this field visit we report key volcanic features and provide some recommendations for future research that could contribute significantly of our understanding of formation of volcanic craters by explosive eruptions. Due to the preliminary nature of this field research here we provide a brief summary of the key geological features identified.

4.3.2 Volcanic Geotopes of Harrat Hutaymah

After an initial pilot project four volcanoes have been identified to have high volcanic geoheritage value and should be part of future geoeducation, geoconservation or geotouristic programs (Moufti et al. 2015). These are the Tabah crater, Harrat Hutaymah crater, Jubb crater (near Ni'ayy village) and the Humayyan/Haram crater (Fig. 4.25).

The **Tabah** crater geotope [27° 1′ 36.34″N; 42° 10′ 6.48″E] is a broad low-rimed volcanic crater about 1.3 km across (Fig. 4.26). It is fairly symmetric and the present day volcanic depression is surrounded by a low crater rim composed of about a few metres to up to 25–35 m thick volcanic succession of pyroclastic rocks (Fig. 4.27).

The crater floor of Tabah is flat and partially filled with reworked voclaniclastic material mixed with some aeolian deposits (Fig. 4.26). The crater wall is steep in the western side where it forms a well-exposed continuous outcrop that starts from the crater floor (Fig. 4.27). The pre-eruptive rocks are not exposed in the crater wall in this site, however in the northern segment of the crater interior granitoid rocks are exhumed that are sitting on a bench-like feature about

Fig. 4.27 Overviews of the Tabah crater [27° 1′ 36.34″N; 42° 10′ 6.48″ E] toward west (**a**), toward south-west (**b**) and toward north-west (**c**). Note the *green* plantation marking the inferred extent of the maar crater on (**a**) and (**c**). Pre-crater scoria and spatter cones (sc1) with lava flow (lf) and a feeder dyke (fd) form an aligned zone that can be connected to another scoria cone on the outer tuff ring flank (sc2). Note the exposed basement rocks on the surface of the erosionally enlarged present day crater floor on (**c**). Tuff ring beds (tr b) are exposed in the inner crater wall showing plastering effect against the steep wall of the crater

few tens of metres above the centre part of the depression (Fig. 4.27). These granitoid rocks are inferred to represent an exhumed syn-eruptive paleo-surface that marks the depositional surface on what the crater rim deposits accumulated. The location of the exposed, in situ granitoid rocks is about 200 metres from the present day crater wall suggesting significant retreat of the crater wall outward causing an apparent enlargement of the crater itself (Fig. 4.27). To constrain the syn-eruptive paleosurface and the crater inner morphology, a combination of geophysical methods, including MT, gravimetry, and geomagnetics need to be conducted in the near future. The centre part of the volcanic depression hosted a village that was well-served by drilled water sources which supported a palm plantation.

Due to water withdrawal the centre part of the depression has gradually subsided and resulted in a critical situation in the 1980s that forced the relocation of the village outside the crater. Today, numerous cracks on preserved buildings and structures are visible on the base of the crater floor where the early settlement was situated (Fig. 4.28). The cracks on the crater floor are restricted to the centre and deepest part of the present day crater, suggesting that the crater that is likely underlain by a diatreme (volcaniclastic sediment-filled volcanic conduit) is located in the centre part of the present day volcanic depression (Blaikie et al. 2014; Valentine and White 2012) and therefore the present day crater wall is an erosional feature and not the original crater wall. Similar crater area enlargement due to erosional processes has been proposed in other unusually large flat-floored maar craters such as those in the Auckland (New Zealand) (Cronin et al. 2009; Németh et al. 2012a) or in the Newer (Australia) Volcanic Fields (Jordan et al. 2013). In this respect the maar craters of Harrat Hutaymah are scientifically important and potentially can serve as excellent examples of an unusual volcanic landscape evolution process. In this respect these volcanoes could support volcanic geoeducational programs that would actively feed from scientific research.

In the southern margin of the volcanic crater a lava spatter cone with clastogenic lava flows and associated feeder dykes have been mapped (Fig. 4.27b). The clastogenic nature of the lava flows (Sumner 1998; Valentine and Gregg 2008) is constrained by the abundance of dark clast outlines preserved in the otherwise coherent solidified lava bodies exposed in the preserved volcanic edifice in the middle of the present day crater (Fig. 4.27). These volcanic rocks stratigraphically underlie the pyroclastic successions forming the tuff ring

Fig. 4.28 Ghost town with date plantation in the Tabah maar crater in Harrat Hutaymah [27° 1′ 49.30″N; 42° 10′ 7.32″E]

surrounding the volcanic depression and are inferred to represent a pre-existing volcanic feature prior to the current volcanic crater (Fig. 4.27). This scoria and lava spatter cone complex is part of a north-south trending chain of at least three volcanic cones that are partially eroded and form a line of about 4 km long in the southern part of the tuff ring.

The tuff ring is nearly intact and there is no characteristic breaching through it other than some narrow gaps that were the pathways for inhabitants lived in the crater (Fig. 4.29). The stratigraphy of the tuff ring-forming succession is fairly uniform across the entire tuff ring, and only some minor variations can be identified that are inferred to reflect variation of transportation axis of pyroclastic density currents, relative distance from the explosion locus and variations in the 3D geometrical position of the tuff ring rim in relationship to the position of the explosive eruption source (Fig. 4.29a). The base of the tuff ring (the base is not exposed) consists of a tuff breccia and lapilli tuff succession that is about 15 m thick in its thickest part (in the eastern and southern quadrant). This stratigraphy unit is rich in accidental lithic fragments and light coloured fine matrix. The base of this unit is more coarse- grained and upward a clear gradual change to a dune bedded coarse-fine-coarse alternation of pyroclastic beds is prominent. The middle section is a dark colour, juvenile ash and lapilli-rich bedded to dune-bedded and cross-bedded dark colour unit with variable thickness (Fig. 4.29a, b). Its thickest part is about 15 m thick in the eastern segment of the tuff ring.

This unit's base composed of a few metres thick succession that is very rich in cored bombs with mantle nodule and megacryst cores. In the upper section the bedding of this unit is well-developed with some spectacular antidune to dune structures, chute-and-pool features, impact sags and numerous evidence for micro-relief and interacting PDC deposit suggesting that this unit is dominated by PDC deposits and its fed from magma probably deeper sourced (e.g. mantle nodule abundance) (Fig. 4.29a, b) making this location a globally unique geotope with numerous excellent geosites. The pyroclastic density current successions are well-exposed and the textural features can be seen in them are in the same quality as those exposed around the Laacher Sea in the Vulkaneifel Global Geopark in Germany (Bogaard and Schmincke 1984; Fisher et al. 1983; Schmincke et al. 1973; Schumacher and Schmincke 1990). The uppermost stratigraphy unit is starts with a light coloured tuff breccia and lapilli tuff that are inferred to be deposited from PDCs (Fig. 4.29a). The section top part is

Fig. 4.29 The thickest tuff ring section is about 30 m thick in the eastern margin of the Tabah crater [27° 1′ 36.34″N; 42° 10′ 6.48″E] (**a**) and exposes at least three major stratigraphy units (u1–3). Around the inferred boundary of u1 and 2 long wavelength dunes and antidunes are exposed indicating a pyroclastic density current origin of the majority of the deposits (**b**, *yellow rectangle* on **a** marks the location shown on **b**). In the western edge of the Tabah crater typical compressed tuff ring succession is exposed that are inferred to represent more distal base surge-dominated pyroclastic succession, however the triplicate stratigraphy still can be recognized. Pre-crater scoria cones of s1 and s2 are also marked on the image. Note the exposed basement rocks in the present day crater basin as well as the reddish tan of the crater floor suggesting the proximity of the sun-eruptive granite surface to the present day crater floor

rich in juvenile ash and lapilli. This stratigraphy can be mapped along the entire volcanic depression however the total thickness of the tuff ring deposits is greater in the east and less in the west. In the western side the preserved tuff ring deposits mimic a medial to distal section (e.g. better bedding, abundance of relatively short wavelength dunes) of a tuff ring suggesting that the present day crater wall (cliff) is an erosional feature and the structural boundary of the original crater might be much closer to the center of the present day depression (Fig. 4.29c). Indeed about 300 m from the present day crater wall on the crater floor granitoid basement rocks crop out that are inferred to represent the syn-eruptive paleosurface (Fig. 4.29c).

Harrat Hutaymah geotope is a maar [26° 59′ 19.39″N; 42° 14′ 50.85″E] with a volcanic depression that is about 120 m deep from the top of its well-preserved tuff ring crest to the crater floor (Fig. 4.30). The crater floor is flat and it hosts a temporal lake that is located slightly in the eastern edge of the crater floor. In the present day crater wall about 50 m above the present day crater floor is a contact between pre-volcanic succession and country rocks of Proterozoic granitoid rocks (Fig. 4.31). The Proterozoic granitoid rocks are covered by about 3–5 m thick siliciclastic deposits (e.g. aeolian and fluvial sand and silt). This siliciclastic succession is covered by at least three distinct lava flow units with lava foot and top breccias each having an average thickness of about 2 metres (Fig. 4.31b–d). This lava flows seem to be tabular and laterally extensive with no systematic thickness variations (Fig. 4.32). This basal lava flows are covered by an about 1–2 m thick aeolian/fluvial deposits (e.g. part is weakly developed soil). In the southern side of the preserved crater a lava spatter-dominated cone form a marked volcanic edifice that is half sectioned in a similar way as it has been identified at Al Wahbah in Harrat Kishb (Fig. 4.31).

This volcano is a complex volcanic feature with at least three distinct half sectioned volcanic craters and a relatively small ponded lava below its main crater exposed in the crater wall. Small lava flows that are inferred to be dominated by clastogenic flows can be traced in the vicinity of this pre-maar scoria- and spatter cone (Fig. 4.32). In the flank of this pre-maar volcanic cone finer grained and bedded, light coloured tuff breccia and lapilli tuff beds mantling the cone edifice. The top of the cone is not covered by these deposits associated with the maar crater formation or there are just thin accidental lithic-rich veneer deposits can be recognized.

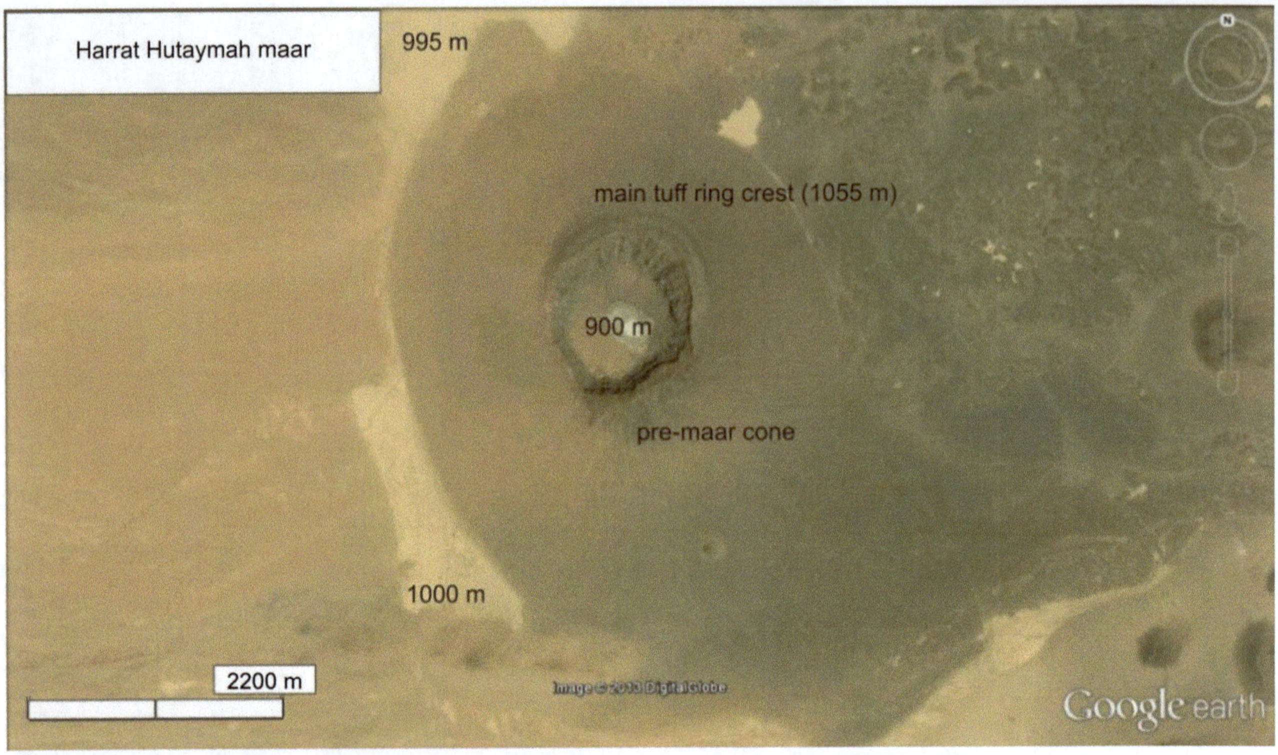

Fig. 4.30 Harrat Hutaymah maar [26° 59′ 15.11″N; 42° 14′ 43.48″E] on a GoogleEarth image. Elevation values refer to the elevation of the outet and the intra-crater surface elevations

The pyroclastic succession inferred to be part of the tuff ring that formed around the maar crater is thicker in the northern side of the crater suggesting a fairly asymmetric volcanic edifice. In the northern flank of the Harrat Hutaymah the tuff ring is steep in the proximity of the crater wall and gradually flattening out toward the north. The edifice can be traced about a kilometre from the present day crater wall. Further away from the crater wall, the bedding become well-developed and the entire succession is composed of dune-bedded lapilli tuff and tuff. The base of the tuff ring is a tuff breccia that is rich in accidental lithic fragments and generally thickly bedded. In the middle stratigraphy position the bedding is improved and fine grained lapilli tuff to tuff beds dominate with various ratio of accidental to juvenile pyroclasts. In this section there are some few-dm thick fall-dominated beds that are rich in cored bombs, and numerous mantle nodules. The uppermost succession of about 25 m thick lapilli tuff is dominated by well-bedded, inverse-to-normal graded juvenile lapilli and ash rich pyroclastic beds that contain cored lapilli and mantle nodules and deep crustal xenoliths (Fig. 4.33). The stratigraphy around the maar is fairly uniform (Fig. 4.33c, d).

Jubb geotope [27° 10′ 51.38″N; 42° 16′ 49.53″E] is a volcanic depression located in the northern part of the Hutaymah Volcanic Field next to Ni'ayy village (Fig. 4.34). It has an abandoned village in its crater similar to Tabah. The villagers were relocated due to an observed gradual crater floor subsidence noticed since the middle of the 80 s as a result of intensive ground water withdrawal (Al-Harthi 1998; Al-Rehaili and Shouman 1985; Bankher and Al-Harthi 1999; Roobol et al. 1985; Vincent 2008). The modern village is currently located in the eastern side of the tuff ring sitting on an alluvial fan (Fig. 4.34). Jubb is a volcanic depression and sits entirely on/between Proterozoic granite land (Fig. 4.34). In the present day volcanic depression's eastern side some granites crop out indicating that this area represents an exhumed syn-eruptive surface on what the tuff ring sits on (Fig. 4.35). This is inferred to be a similar situation to those identified at Tabah crater but in a much clearer 3D view. The tuff ring is thicker in the eastern side of the volcanic depression—up to 40 m—while in the west the crater rim is about 20 m thick in total (Fig. 4.35).

The tuff ring stratigraphy composed of three major units (Fig. 4.36a, b). The basal unit is dominated by accidental lithic fragments from various shallow and deep sourced country rocks, mantle nodules and angular juvenile fragments. Fluidal shape juvenile bombs are usually large and cored with various country rocks. The same light coloured basal unit exposed also in the western side of the crater but it is finer grained, rich in accretionary lapilli, vesicular tuffs and individual ballistic bombs commonly form impact sags on underlying beds (Fig. 4.36c). The middle section is more

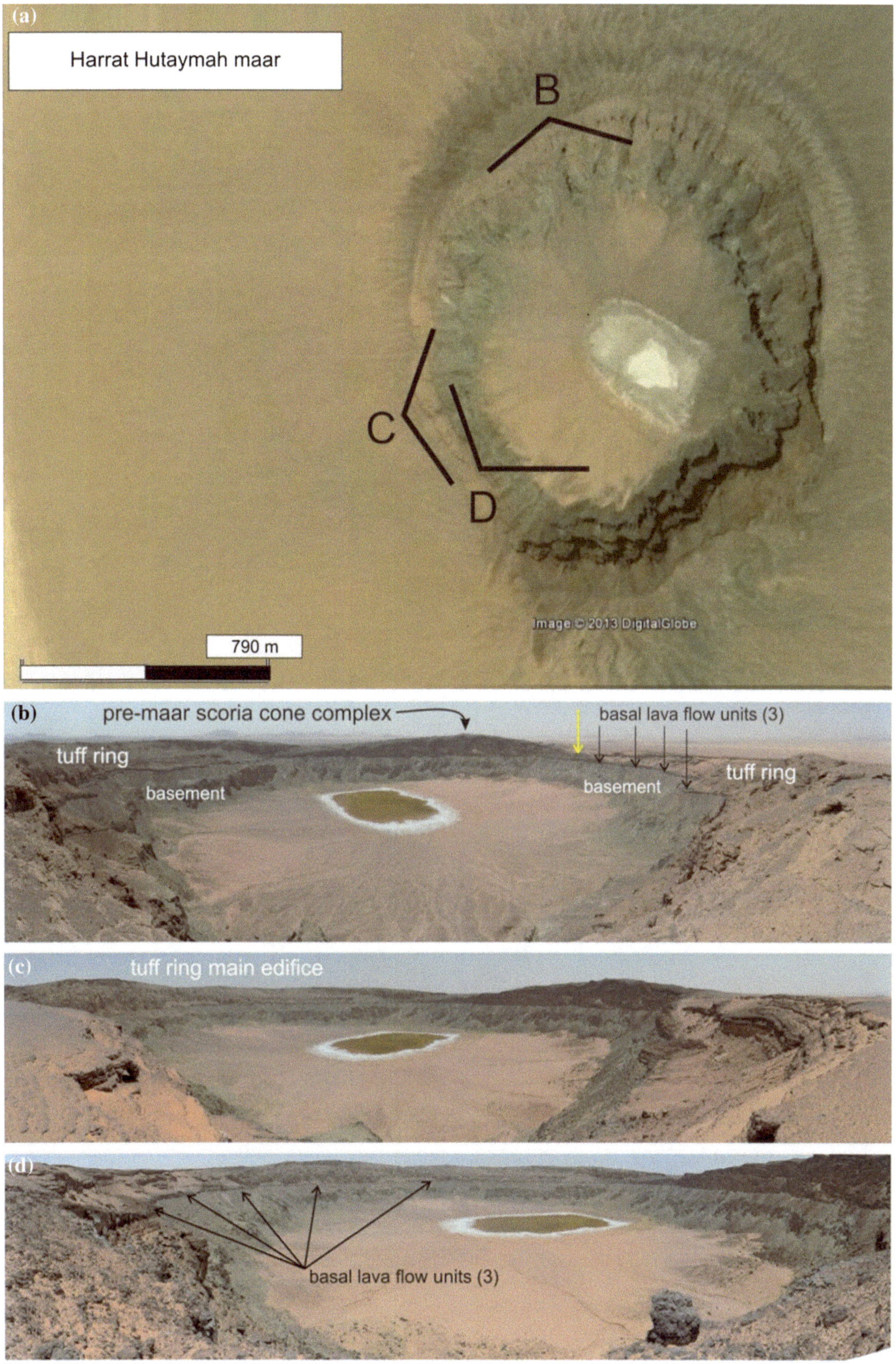

Fig. 4.31 *Close up view* of Harrat Hutaymah maar on a google earth image (**a**). Three separate *panoramic images* show the key features of the Harrat Hutaymah [26° 59′ 15.11″N; 42° 14′ 43.48″E] crater (**b–d**). *Yellow arrow* on **b** points to a lava flow initiated from the pre-maar scoria cone

Fig. 4.32 Basal pre-crater lava flows exposed in the crater wall of Harrat Hutaymah volcano [26° 59′ 10.97″N; 42° 14′ 29.54″E]. The basal lava flows consist of at least 3 lava flow units and they are sitting over a siliciclastic sedimentary succession estimated to be at least 5 m thick over the granitoid basement rocks

Fig. 4.33 Cored bombs with mantle nodules in the upper pyroclastic units of Harrat Hutaymah [26° 59′ 15.11″N; 42° 14′ 43.48″E] maar (**a**). Intact cored bomb with contractional cracks (**b**) are signs of fast chilling of magma upon fragmentation (**b**). Antidunes (**c**) and matrix supported base surge beds (**d**) dominates the Harrat Hutaymah tuff ring succession

juvenile pyroclast dominated and exceptionally well bedded in its middle part with some spectacular moderate wavelength (0.5–3 m) antidunes (Fig. 4.37a, b). Antidunes are exceptionally well-exposed in the road cuts enter to the crater from outside (Fig. 4.37c, d).

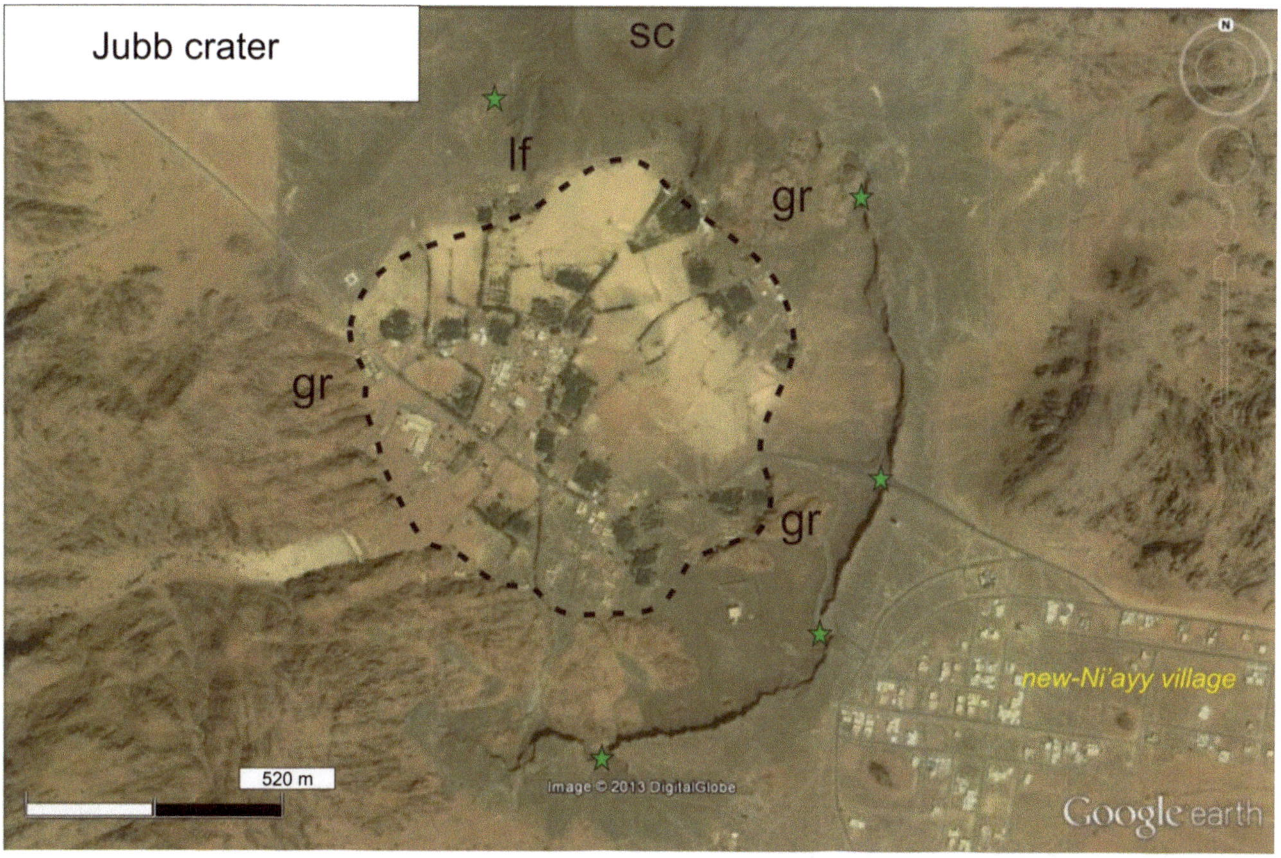

Fig. 4.34 Google earth image of the Jubb crater near new-Ni'ayy village [27° 10′ 25.28″N; 42° 17′ 33.45″E]. The present day volcanic depression is an erosionally enlarged volcanic landform. The structural boundary ot the original maar crater is inferred to be located in the area (*dashed line*) where the crater floor is relatively flat and filled with various sediments (loess, silt, sand etc.). Exhumed basement granitoid rocks are marked by *gr*. *Green stars* are measured sections. A pre-maar scoria cone is labelled by *sc* while a small lava flow marked by *lf*

The topmost succession is more abundant in juvenile fragments, and an increase of fusiform lapilli is prominent in the uppermost part of the succession that can be defined as a separate unit (Fig. 4.37e). In the northern side of the volcanic depression an aligned lava spatter/scoria cones represents a pre-eruptive morphology. This volcanic cone complex has some basal lava flows that are exposed in the present day crater. The crater-ward side of the cone truncated by landslides but it has not been half-sectioned as it the case in Harrat Huttaymah. This indicates that the structural boundary of the volcanic crater must be closer to the center of the present day depression and this crater has also been erosionally enlarged similarly as it has been inferred for the Tabah crater.

The Jubb is a truly complex volcanic geotope with high geoheritage value. The preserved pyroclastic successions show a complex explosive eruption story that were violent and energetic similar to well-known sites in Europe like the Laacher Sea in Germany. The advantage of exposures at Jubb however is that the arid climate has a unique effect on the preservation potential of the pyroclastic rocks providing fantastic exposed 3D volcanic facies architecture to see in a confined and relatively easy to access region. The fact that a small township and roads are crossing the maar crater made this location as a perfect site for geoeducational and geotouristic programs.

Humayyan/Hamrah volcano [27° 11′ 4.46″N; 42° 22′ 45.07″E] is a complex volcanic geotope that is located about 10 km to the east from Jubb. It consists of a large crater-like depression that is located in a centre of an extensive lava flow field. In the crater wall no pyroclastic rocks exposed, suggesting that this depression might be a pit crater formed on a top of a growing lava shield (Fig. 4.38).

In the southern part of the volcanic complex however exposes thick succession of pyroclastic rocks rich in accidental lithic fragments, mantle nodules, and abundant angular volcanic pyroclasts hosted in a fine ash matrix. This pyroclastic succession is clearly covered by the lava flows initiated from a lava shield host a large (double) pit crater (Fig. 4.38). This facies relationship suggests that a large tuff ring must have been formed prior the lava shield and subsequent pit crater formation. In a large single horst in the SE exposes about 100 m total thickness of tuff breccias and

Fig. 4.35 Panoramic views to Jubb crater [27° 10′ 47.34″N; 42° 16′ 49.10″E] toward NW (**a**), to the E (**b**) and toward the SW (**c**). Note the exhumed granite surfaces (*gr*) in the interior of the present day volcanic depression draped by base surge deposit dominated section of a tuff ring (*tr*)

lapilli tuffs (Fig. 4.39). In this section a threefold stratigraphy can be identified (Fig. 4.40). The basal succession is a grey tuff breccia and lapilli tuff that is stratified, massive and contains abundant angular juvenile fragments as well as lithic pyroclasts. This about 60 m thick unit is covered by a yellow tuff breccia (~20–25 m thick) that is in angular unconformity with the basal grey pyroclastic unit and forming a dish-filling nature toward the centre of the volcanic complex. This yellow pyrolcastic unit contains large accidental lithics that are over 4 m in diameter. The top of the succession composed of an about 5–10 m thick matrix-supported units that contains bed-flattened lava spatters and agglutinated pyroclast horizons. In the center part of the volcanic complex this basal pyroclasts clearly covered by lava flows that can be correlated around the depression. In the southern part of the entire volcanic complex along an entry path to the main pit crater the present day morphology is subdued by strong erosion and exposing individual buttes with a very similar stratigraphy but various bedding orientation and attitude suggesting that this zone might be the former crater of a tuff ring.

For reconnaissance purposes the expedition visited another tuff ring that is completely filled with aeolian deposits in its former crater. This volcano also exposes a nice section of typical tuff ring succession with accidental lithic rich lapilli tuffs and tuffs that hosts accretionary lapilli and numerous angular ash and fine lapilli consistent with an explosive phreatomagmatic origin. In addition scoria cones forming the central chain of the volcanic field suggest their NS aligned nature, fissure-like distribution is manifested in the individual volcanic edifice natures.

4.3.3 Main Findings and Geoheritage Value of Harrat Hutaymah

Four of the most prominent and best exposed volcanic explosion craters of the Harrat Hutaymah inferred to be

Fig. 4.36 Pyroclastic units (u1–3) of the Jubb crater [27° 11′ 1.83″N; 42° 17′ 17.68″E] form thick tuff ring pile in the eastern edge of the volcano (**a** and **b**), while in the western edge pyroclastic rocks show textures and bedding characteristics are more typical for distal PDC units. In the northern side of the crater the tuff ring pyroclastics sit on a pre-crater scoria cone (*sc*) and associated lava flow (*lf*). Granitoid basement rocks are exposed in the erosionally enlarged maar crater (*gr*)

formed due to phreatomagmatic explosive eruptions. In each of the detailed studied volcanoes it can be conclude that their formation was dominated by magma fragmentation that produced chilled pyroclasts as well as fine ash and excavated country rocks from various levels. Accretionary lapilli and vesicular tuffs were recognized in Tabah, Hutaymah and Jubb volcano, especially in sections typical for medial or distal part of a tuff ring. At Hamrah however, the exposed (and visited) sections are likely representing proximal tuff ring successions and therefore such features are not expected to be seen. However, at Hamrah the tuff ring forming successions are typical for explosive eruptions produce abundant country rocks, deep seated xenoliths and transported through relatively cold PDCs that are consistent for a proximal succession of tuff rings formed due to phreatomagmatic explosive eruptions.

From the four identified geotopes and their geosites Harrat Hutaymah is the best preserved and it is likely the most intact volcano. The fact that country rocks are exposed in the crater wall well above of the crater floor clearly classify Harrat Hutaymah to be a maar volcano (White and Ross 2011). In the case of Tabah and Jubb the volcanic edifices are more subdued and the erosion likely formed a significant retreat of the crater wall toward the distal part of the tuff ring resulting exhumation of syn-eruptive surfaces on the granitoid landscape. In addition these two craters are also filled nearly completely, and to identify the structural boundary of the craters is difficult and potentially a subject of excellent geophysical surveys planned in the future as it has recently been demonstrated from other low profile wide craters from the Newer Volcanics in Victoria, Australia (Blaikie et al. 2012). Recent researches also targeted flat, broad maars with an aim to determine the number of eruptive sites, and the role of vent migration across a broad crater area to form large amalgamated maar craters versus the fact how wave-cut erosion and lake infill processes can enlarge an original maar crater (Boyce 2013; Jordan et al. 2013; Németh et al. 2012a). The broad volcanic craters of Harrat Hutaymah are the perfect sites to contribute to this frontline research questions.

At this stage it can be said that these two volcanic depressions are also formed due to explosive phreatomagmatic eruptions and they are also maar volcanoes. In case of Humayyan/Hamrah, the present day large volcanic depression is inferred to be a pit crater complex on a large lava shield. However, the initial eruption of the Humayyan/Hamrah must have also been explosive phreatomagmatic and produced a fairly large and extensive phreatomagmatic volcano once occupied the southern part of the volcanic complex. In many respect the lower part of the exposed

Fig. 4.37 *Panoramic views* of the tuff ring succession [27° 10′ 41.86″N; 42° 17′ 17.88″E] in the eastern sector of the tuff ring of Jubb with its pyroclastic units (u1, u2, u3 and u4) (**a–c**). On **c** and **d** *yellow circles* locate the well-developed dunes/anti-dunes of the PDC deposit

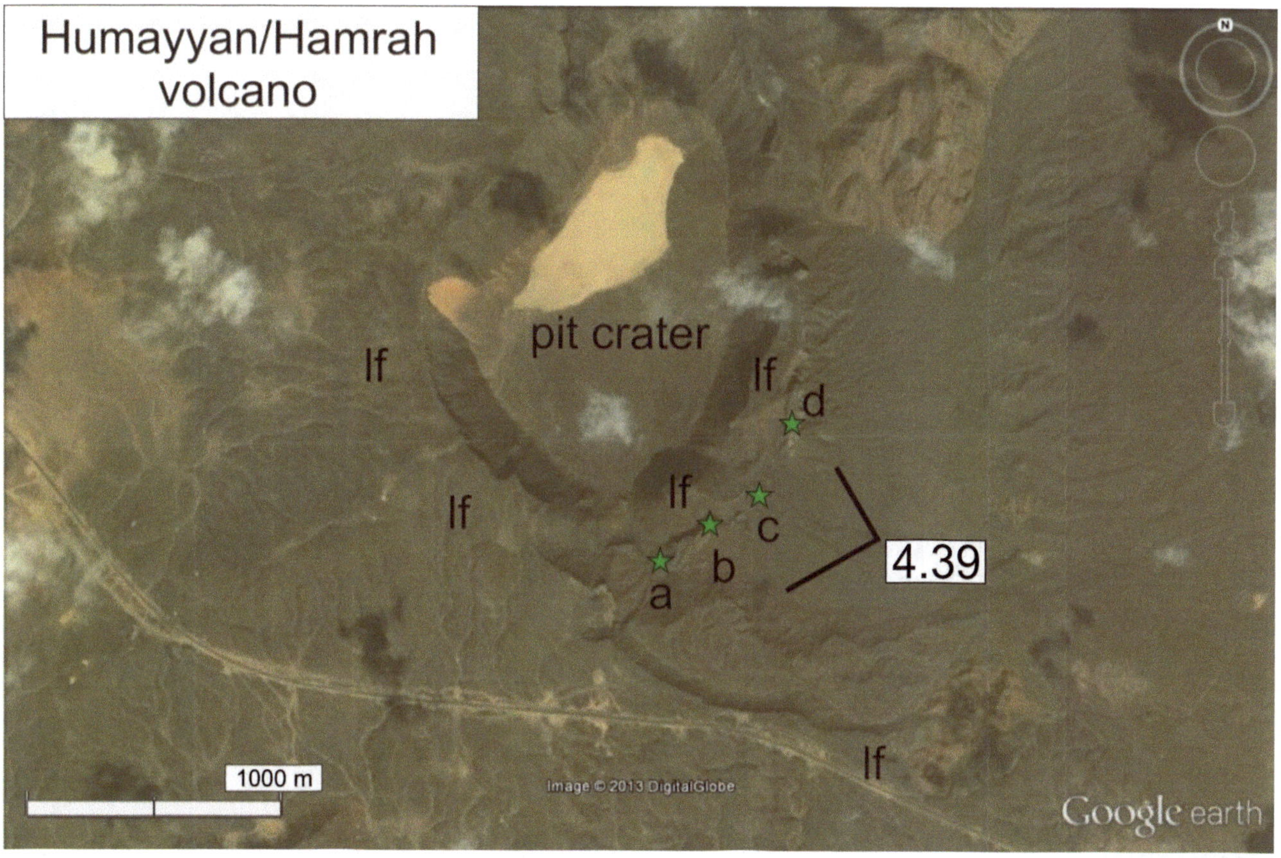

Fig. 4.38 Google earth image of the Humayyan/Hamrah volcano with its large pit crater [27° 11′ 6.57″N; 42° 22′ 33.59″E]. Pyroclastic rocks of part of a tuff ring exposed under lava flow (*lf*) units (part of a lava shield) in the SE sector of the volcanic complex. View-point from where the Fig. 4.39 was taken is marked on the map. Signs from **a** to **d** represents large cliffs composed of pyroclastic rocks, a dissected part of a former tuff ring

Fig. 4.39 View toward the NW to the main crater of Humayyan/Hamrah volcano [27° 10′ 39.54″N; 42° 22′ 56.00″E]. *Yellow line* marks contact between lava flow units and pyroclastic successions (tuff ring - *tr* and its units u1 and u2). *White line* marks the contact between phreatomagmatic successions and capping lava spatter-dominated tuff breccia

pyroclastic units resemble textural features typical to diatreme filling successions similar as it has been reported from Hopi Buttes, Arizona (Lefebvre et al. 2013; White 1991), Waipiata, New Zealand (Németh and White 2003), or the Pannonian Basin, mostly in Hungary (Németh et al. 2001). In this respect the exposed pyroclastic rocks could represent an exposure level of volcanoes near or even below the syn-eruptive surface. To map the specific units and correlate them it would help to clarify the 3D architecture of this interesting volcano. It is however, can be inferred, that the eruption mechanism to form this volcano was also phreatomagmatic and produced significant volume of pyroclasts that transported and deposited by various particle concentration PDCs. Overall, the visited sites confirmed that at Hutaymah Volcanic Field phreatomagmatism and in general explosive volcanism that produced primarily PDCs were a major eruption style, which is indeed makes this field outstanding from other harrats.

Fig. 4.40 Composite views of the main pyroclastic succession of the preserved tuff ring in the SE sector of the Humayyan/Hamrah volcano [27° 10′ 44.09″N; 42° 22′ 50.20″E]. Three major stratigraphic units (u1, u2 and u3) could be distinguished in locations about a kilometre apart from each other. On **b** people for scale marked in a *circle*

The volcanic explosion craters of Hutaymah Volcanic Field can provide an excellent playground for modern geophysical techniques to conduct including gravimetry, MT and geomagnetic survey to delineate the structural boundaries and the nature of the crater filling rocks. In combination with modern geochemical and geochronological work, these studies could yield to a state-of-art research that could significantly contribute to current cutting edge research on crater formation processes. With this high scientific potential these region can generate high scientific interest that can build a strong scientific foundation over years to describe, catalogue and rank volcanic geoheritage value in the region. The good exposures, the easy access and the high aesthetic value in addition made this region an ideal location to develop geoheritage projects that may culminate in the formulation of regional or global geopark in the near future.

4.4 Harrat Khaybar

4.4.1 Overview

A reconnaissance visit to Harrat Khaybar (Fig. 4.41) was performed recently (in 2013) with an aim to collect information on the general field conditions, the accessibility of the key geological sites and asses the geoheritage value of the region by identify key geological sites that could be used in the future for detailed geological projects. The southernmost part of the field was explored where potential historic eruptions sites are suspected as well as various silicic tuff rings are located. In the central part of the field geology work was concentrated on the Jabal Quidr region as this volcano being the youngest of the region as reported and generally accepted (Camp et al. 1991; Chagarlamudi et al. 1991; Coleman and Gregory 1983; Demange et al. 1983). In addition this volcano shows many very young volcanic features such as lava flow surface textures and lack of erosional features on its flanks. This volcano also associated with an extensive ash plain that can be traced at least 17 km from its source (Fig. 4.29), putting this volcano and the eruption produced this ash plain into the eruption range of being sub-Plinian or violent Strombolian (Valentine and Gregg 2008). Near Jabal Quidr, dual silicic volcanic systems were the main interest of the field visit. The so called "*White Mountains*", Jabal Bayda and Jabal Abyad are comenditic centers (Baker et al. 1973; Camp et al. 1991; Demange et al. 1983) produced a tuff ring and a lava dome complex with short run out distance block and ash flow deposits and associated obsidian lava domes and coulees. Just north of Jabal Quidr an extensive fissure zone with pit craters and a chain of lava spatter cones were visited to study lava flow surface textures and their implications to lava flow rheology and behavior. In addition the region exceptional geoheritage value was also documented.

The dominant rock types of the western Arabian Miocene to Recent intracontinental volcanic fields are hawaiite, but subordinate, more felsic rock types, such as benmoreite, mugearite and trachyte, are also known, especially in the largest volcanic fields with the most complex volcanic stratigraphy, such as the Harrat Khaybar (Camp et al. 1991). Harrat Khaybar basal volcanics formed the Jarad Basalt (5–3 Ma) that is overlain by the Murash Basalt (3–1 Ma) and it is capped by the Abyad Basalt (1 Ma—Recent) (Camp et al. 1991). Harrat Khaybar has the most prominent felsic volcanoes of the Arabian Peninsula erupted from a compositionally zoned near-surface magma chamber along a N-S fault zone in the central part of the field (Camp et al. 1991).

4.4.2 Jabal Quidr [25° 43′ 11.23″N; 39° 56′ 37.32″E]

Jabal Quidr is one of the most prominent volcanic landform of the Harrat Khaybar (Fig. 4.42). It dominates the center part of the volcanic field with its near perfect symmetric volcanic cone that is composed of a basal gentle sloping lava flow dominated part and crowned by a reddish volcanic cone (Fig. 4.42a). The slope angle changes coincide well with the boundary between a basal lava shield and a capping volcanic cone that has formed due to explosive volcanic eruption and accompanied crater subsidence triggered by lateral drainage of a central crater on top of it (Fig. 4.42b). The base of the volcanic edifice composed of complex pahoehoe lava flows that characterised by few m wide lava tubes that cross cut each other forming a complex network of tube-fed solidified lava fields. In major axial zones, uplifted lava crusts commonly twisted and rotated forming several tens of metres wide zones of inflated and deflated lava ponds commonly associated with lava tumuli (Anderson et al. 2012; Duraiswami et al. 2004) (Fig. 4.42a). Individual lava tubes are partially covered and connected with zones of outflows where cm to dm scale lava fingers form a complex surface texture (Fig. 4.42b).

Larger inflated lava flows are commonly forming blistery surfaces with shelly pahoehoe surface textures (Stevenson et al. 2012) (Fig. 4.43a). These lava blisters normally in a few metres across size and in many occasions their roof is collapsed and broken, exposing large voids beneath them (Fig. 4.43a). In other cases, especially along the base of the volcanic edifice lava flows are commonly ponded and thick lava crusts exposed along fractured and uplifted/downthrown margins of tumuli (Fig. 4.43b). Complex surface wrinkle textures (Fig. 4.43c, d) suggest some outpouring of fresh lava along cracks on the lava tubes and ponded surfaces that were then subsequently sheared away. These surface textures suggests that lava flows from Jabal Quidr were low viscosity and moved fast in the high slope angle regions preventing to form extensive and thick lava crusts commonly broke apart

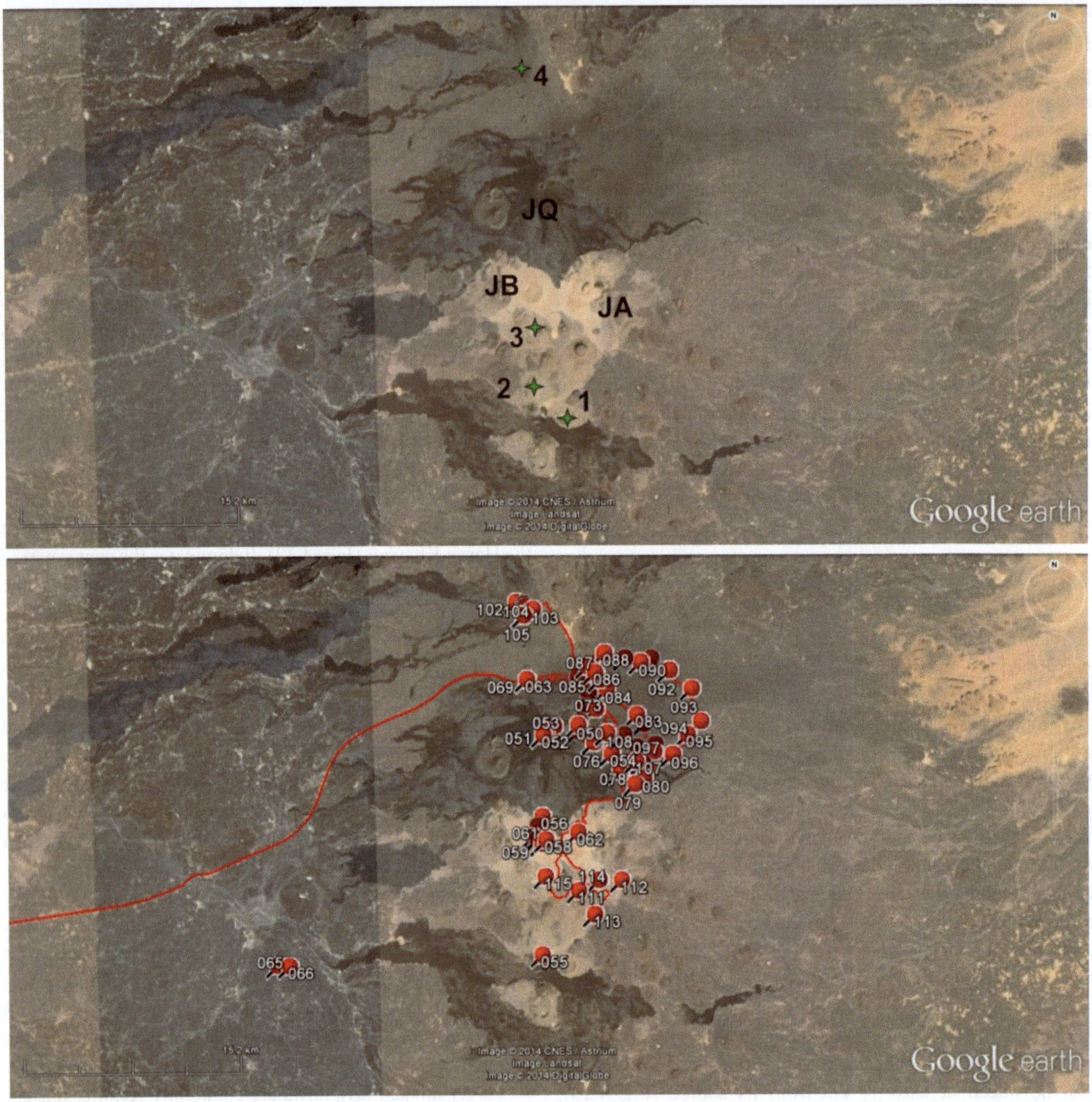

Fig. 4.41 Harrat Khayber in Google earth satellite images. The *upper image* shows the key locations visited during the field campaign: *JQ* Jebel Qidr, *JB* Jebel Bayda, *JA* Jebel Abyad, *1* young silicic tuff ring with broad crater, *2* inverted silicic tuff ring with eroded rims, *3* well-preserved complex silicic tuff ring with broad crater, *4* lava spatter complex with pit crater network and extensive lava outbreak points. The bottom immage shows the waypoints and tracks coverred during the field campaign. 065–066 are the locations of the entry points of a major lava tube network

the freshly ponded crusts while in the low slope angle regions where lava flow movement slowed down ponding and crust formation were more pronounced (Keszthelyi and Denlinger 1996). The common mechanical stress on large lava crusts indicates repeated inflation and deflation processes in the ponded lava zones (Calvari and Pinkerton 1999; Hoblitt et al. 2012; James et al. 2012) that is inferred to be controlled by the inflow and outflow of melt from the ponded zones. The lava flow surface texture variations and the common development of transitional lava fields has been documented as a widespread feature of the Harrat Rahat (Murcia et al. 2014) that is seemingly the case at the Harrat Khaybar as well.

In the feet of Jabal Quidr large ponded lava flows form thick crusted ponded lava zones where the crust is over 1 m

Fig. 4.42 **a** Overview of Jebel Qidr from the east [25° 43′ 36.74″N; 39° 57′ 4.39″E]. Note the shape of the cone having a well-distinguished scoria cone over a shield-like edifice. **b** The gentle dipping slope of the eastern flank of Jebel Qidr is dominated by pahoehoe lava flows and abundant tumuli and pressure ridges. The *upper section* of the lava flows near the initiation points (*boccas*) lava flows are rubble demonstrating fast removal of freshly developed lava crusts while in the *lower section* of the edifice flank flow movement might have been less vigorous allowing to develop proper lava clasts and fantastic pahoehoe surface textures

thick indicating relatively stable conditions of the lava ponds to develop thick crust (Fig. 4.44a). On the basis of the texture and the thickness of the lava crust it can be inferred that the time needed to form such crusts is in the range of days to weeks suggesting a relatively stable melt supply to keep this lava ponds stationary over long time. In places where new melt entered to a ponded lava zone an open roof channel might formed along the lava moved and commonly changed its level as it is evidenced from solidified open channel systems commonly associated with local tumuli (Harris et al. 2009; Patrick and Orr 2012; Stovall et al. 2009) (Fig. 4.44b).

The volcanic edifice of Jabal Quidr is primarily composed of lava spatter dominated breccias and agglomerates interbedded with welded spatter, clastogenic lava flows and relatively thin (m-scale) fluidal lava flows covering the outer edifice rim in a sheet-like fashion (Fig. 4.45). The agglomeratic pyroclastic breccia is mixed with vesicular pyroclasts of red, brown and balck lapilli and ash (Fig. 4.45a).

Fig. 4.43 Lava surface textures from the proximal lava flow regions of the Jebel Qidr volcano [25° 43′ 41.34″N; 39° 57′ 45.68″E]. Note the thin and thick lava crusts develop two different types of lava tube network (**a** and **b** with *white circle*). Common features are the ropy basalt textures and wrinkles on an uplifted lava tube roof commonly associated with break out tumuli in the ponded lower section of the lava flow fields (**c** and **d**)

Occasionally exhalation marks present as evidence of high temperature fumarola and/or solfatara activity after the formation of the edifice. Some cracks, fractures are common feature along the lip of the crater indicating some mechanical instability of the crater that is a common feature on such edifices (Németh et al. 2003; Thordarson and Self 1993).

The inner crater wall of Jabal Quidr is perpendicular (Fig. 4.45b) and exposes a succession of lava flows that are partially covered by drained back lava lounges as an evidence of some lava lake presence in various stages of the crater evolution (Fig. 4.45a). In the cross-section of the lava flow networks there are U-shaped lava ponds exposed indicating the presence of lava lakes in a smaller scale (tens of metres) and subsequent pit crater formation similar to those documented on large mafic volcanoes such as on those in Ambrym (Németh and Cronin 2008) or Piton de la Fournaise (Carter et al. 2007). The size of the crater is about 600 m across which is a large size capable to host significant volume of melt that then later on can be released through flank "boccas" as it has been documented in various outbreak fractures mostly in the western side of the upper outer flank of the cone (Fig. 4.45c). In the western side of the edifice there are in situ lava spatter sections that are steep (over 40 degrees slope angle), and agglutinated together to be erosion resistant remnants, but they also mark a situation if the magmatic pressure behind an edifice increasing, such edifice could become instable very quickly and initiate collapsing sections through lava lakes can be drained in a catastrophic way (Head and Wilson 1989). Similar steep spatter cones were documented from the Harrat Rahat and defined among the steepest on Earth (Moufti et al. 2013b).

Jabal Quidr has been surrounded by an extensive ash plain with a dispersal axis toward NE. The ash plain composed of scoriaceous ash and lapilli beds in several units indicating various episodes of violent explosive activity that were able to provide sustained eruption column and allow pyroclast to be transported beyond 15 km from the source making these eruptions in a range to be sub-Plinian or violent-Strombolian. The ash and lapilli are equi-dimensional to flat but highly vesicular suggesting full expansion of pyroclasts through fragmentation. The ash plain provided a base for the majority of the long lava flows initiated from Jebel Quidr, suggesting

Fig. 4.44 Thick lava crust in the ponded pahoehoe lava flow field of Jebel Qidr's foothill [25° 43′ 41.34″N; 39° 57′ 45.68″E] where the slope angle of the volcanic edifice has changed leading to slow the lava flow down allowing some degree of ponding and thickening thus developing thick crust. The over a m thick crust can be interpreted several days of relatively stationary lava pond to exist in the foothil of the volcanic edifice

that the main violent explosive phase of the eruptions preceded the lava effusion stage. The lava flows reached over 10 km in length and followed morphological depressions, fluvial networks (Fig. 4.46a). In the medial section the lava flows are commonly transitional in type suggesting that they were derived from a tube fed lava that were mechanically abraded and pushed ahead to the medial distance. Where the terrain became relatively flat, the transitional lava flows were able to retain heat enough to maintain some tube to be active and feed further pahoehoe lava flow sheets that spread in a relatively thin skin-like manner over the ash plains (Fig. 4.46b, c). Lava flows were commonly flow into pre-existing craters filling them completely (Fig. 4.46a).

4.4.3 Jabal Abyad [25° 39′ 34.35″N; 39° 58′ 16.50″E]

The "White Mountains" of Harrat Khaybar are strikingly different in their appearance and their eruption history to the most landscape-dominant volcano of the field, the Jabal Quidr, which is a hawaiite stratovolcano and believed to have erupted in historic time and emitted dark lava fields (Camp et al. 1991) that are banked against the white comenditic ash and lapilli plain associated with the "White Mountains" (Fig. 4.47). The "White Mountains" refer to a pair of comenditic volcanoes (Baker et al. 1973; Camp et al. 1991): Jabal Abyad and Jabal Bayda (Fig. 4.47). Both Arabic names mean "White Mountain", with Abyad a masculine and Bayda a feminine form of white in Arabic reflecting that Jabal Bayda is a near perfect circular tuff ring with a shallow crater, while Jabal Abyad is a lava dome complex that forms a hill standing about 300 m above the surroundings (Fig. 4.47). Jabal Abyad is the highest volcano of Harrat Khaybar, reaching 2093 m above sea level, while Jabal Bayda is 1913 m high (Figs. 4.47 and 4.48). Their age is poorly constrained, but inferred to be between 0.86 and 0.22 My (Camp et al. 1991). While felsic lava domes and tuff rings exist elsewhere (Austin-Erickson et al. 2008, 2011; Cano-Cruz and Carrasco-Nunez 2008; Németh et al. 2012b; Riggs and Carrasco-Nunez 2004), the significance of the felsic intracontinental volcanism of the Arabian Peninsula is

Fig. 4.45 *Panoramic views* of the deep crater of Jebel Qidr [25° 43′ 11.23″N; 39° 56′37.32″E]. Note the lava spatter-dominated pyroclastic succession forming a steep edifice on top of Jebel Qidr (**a**). In the crater wall a succession of lava flows of 1–3 m thick exposed with some evidences to infer the existence of lava lakes and pond that subsequently provided to form pit craters upon their release from the base of the edifice (**b**). The steep nature of the edifice is the result of a welded lava spatter-dominated capping units forming the *top* of Jebel Quidr (**c**). Note the slight depression and darker colour zone in the right hand side of the image that is a lava flow outbreak along spatter and lava flow pieces were cascading down on the steep edifice slope

great in terms of understanding the evolution of dispersed magma in near–surface compositionally zoned magma chambers (Camp et al. 1991).

The "White Mountains" are composed of comenditic pyroclastic successions of intercalated small-volume block-and-ash flow, pyroclastic density current and minor air fall units (Figs. 4.49 and 4.50), comenditic lavas including short, but thick obsidian lava flows (Fig. 4.51) and that form very distinct volcanic landforms with white and beige-to-orange colours, making them stand out from the otherwise dark hawaiite, mugearite and benmoreite lava flows, domes, and dome coulees. A recently initiated Arabia Geoparks Project has demonstrated the high geoeducational value of these volcanic landforms of western Arabia and how these could be utilized to understand volcanic hazards and geoconservation (Moufti and Németh 2013). The "White Mountains of Harrat Khaybar" will be flagship geotopes with numerous geosites in the provisional volcanic geopark of the region.

Jabal Abyad is a fantastic volcanic geotope. It has some excellent outcrop to study the short run-out distance block-and-ash flow deposits commonly related with lava dome growth and repeated explosive eruption producing typical pyroclastic density current dominated successions

Fig. 4.46 Distal lava flows of Jebel Qidr [25° 43′ 11.23″N; 39° 56′ 37.32″E] are thin pahoehoe flows along main streamline of lava flows slabby to rubble pahoehoe textures are common. Lava flow fields are extensive and commonly filling gaps between volcanic edifices or completely infil preexisting tuff rings sucha s shown on **a** with *arrow* (*JQ* Jebel Qidr, *JA* Jebel Abyad). Lava flows are commonly less than a meter thick about 7 km from their sources **b** where the lava flows are accumulated over thin ash plain deposits. Truncated surface textures **c** are commonly present in areas where the thin lava flows are blocked against obstacles, or banked agains gently upward sloping landscape

Fig. 4.47 *Panoramic views* show the White Mountains (Jebel Abyad and Jebel Bayda) from the *top* of the Jebel Qidr [25° 43′ 11.23″N; 39° 56′ 37.32″E]. Note the extensive lava fields sourced from Jebel Quidr in the foreground

rich in flow banded lava dome-derived clasts (Fig. 4.50). The numerous individual block-and-ash flow deposits identified in the flank of the Jabal Abyad cone indicate repeated lava dome growth and collapse event through the evolution of the volcano. Some flow banded obsidian unit is still preserved in the middle section of the volcanic edifice (Fig. 4.51). The facies architectue, the relatively small size of the edifice and the small volume of the individual pyroclastic successions made Jabal Abyad comparable in size to those rhyolitic domes commonly form fields in intracontinental settings (Riggs and Carrasco-Nunez 2004).

4.4.4 Jabal Bayda [25° 39′ 38.15″N; 39° 56′ 0.27″E]

Jabal Bayda is a perfectly preserved silicic tuff ring near the Jabal Abyad. It composed of white pumiceous ash and lapilli beds that have some degree of gully network formed on its outer edifice flank (Fig. 4.52a). The tuff ring crater rim is in an even elevation, and forming a relatively broad crater lip. The crater is partially filled with reworked (fluvial and Aeolian reworking) ash and dust, that cut by a gully network sometimes reaching 5 m in depth in their deepest points

Fig. 4.48 Jebel Abyad [25° 39′ 34.35″N; 39° 58′ 16.50″E] from the top of Jebel Bayda [25° 39′ 29.66″N; 39° 56′ 9.22″E]. *View* on **b** is slightly more focused from a slightly different angle than it is on **a**. Note the extensive white debris fan and fluvial network around the lava dome of Jebel Abyad. The hard rocks crop out halfway on the lava dome are interbedded obsidian lava flows and block and ash flow deposits. White pyroclastic density current deposits from Jebel Bayda overlain short run out block and ash flow deposits from jebel Abyad. Note on **a** other steep lava domes in the region dominated by comenditic voclanism both explosive and effusive, lava dome-forming styles

Fig. 4.49 Pyroclastic density current (PDC) deposits in the feet of the Jebel Abyad [25° 39′ 49.60″N; 39° 57′ 36.74″E] showing low angle cross bedding, grading, and relatively unsorted nature. In the middle section obsidial lava flow crops out (OBS)

(Fig. 4.52b). The crater hosts a double lava dome that occupies its northern inner sector. In the SE side of the tuff ring, the highest point of the tuff ring rim is defined by another lava dome that is partially exposed in the inner crater wall and directly connected to proximal block-and-ash flow deposits flanking into the crater basin. In the NW outer edifice flank about half way to the top deep gullies exposes a pyroclastic density current dominated succession in the main pyroclastic facies of the Jabal Bayda tuff ring. The pyroclasts are normally angular, microvesicular, and coated by white siliceous dust. The matrix of the pyroclastic beds is fine silicic ash. Obsidian coarse ash and lapilli as well as silicic volcanic lithic fragments are common. In fine beds ash aggregates can be inferred to be result of pyroclast accretion and they can be defined as accretionary lapilli. Cross bedding, dune bedding and some cross lamination is prominent

Fig. 4.50 Oucrop features of the pyroclastic successions of the Jebel Abyad volcano [25° 39′ 48.84″N 39° 57′ 43.65″E]. Large banded obsidian clasts hosted in fine matric (**a**) forming massive facies commonly underlain by bedded basal layer (**b**). The block and ash flow deposits are matrix supported and unsorted with abundant obsidian clasts (**c**) forming about one meter thick units (**d**)

Fig. 4.51 Flow banded obsidian lava flow in the Jebel Abyad [25° 39′ 40.75″N; 39° 58′ 4.09″E]

Fig. 4.52 Jebel Bayda [25° 39′ 38.15″N; 39° 56′ 0.27″E] is a comenditic tuff ring with a tuff ring of over 50 m high (**a**) that surrpounds a flat floored, broad crater (**b**). Well developed gully network is visible in the outer flank of the tuff ring (**a**), while in the crater a small lava dome is preserved (**b**)

between coarser grained massive, unsorted matrix supported dm-thick beds that are best to interpret as pyroclastic surge beds intercalated and/or associated with small-volume block-and-ash flow deposits. Such pyroclastic architecture is similar as described from other silicic tuff rings elsewhere (Austin-Erickson 2007; Campos Venuti and Rossi 1996; Druitt et al. 1995; Tait et al. 2009).

4.4.5 Other Silicic Volcanoes

Harrat Khaybar hosts several other unique silicic volcanoes that were visited for an initial field work during 2013 November. In the southern edge of the volcanic field several intact tuff rings are preserved such as those numbered as Tr1 (Fig. 4.53a). This tuff rings are similar in size and morphological appearance to the Jabal Bayda, being either perfectly preserved ring structures with various level of gully network developed on their outer edifice flank (Fig. 4.53a) or being partially eroded by rock fall and undercut erosion where their crater is partially preserved, but their proximal edifice eroded and cut back significantly (Fig. 4.53b). From the southern edge of the field toward Jabal Bayda at least 3 major tuff ring complexes are preserved (Fig. 4.53b).

In addition to the silicic tuff rings in the southernmost part of Harrat Khaybar an apparently young, and potentially historic eruption site is evident (Fig. 4.54a, b). A black, steep scoria cone form a prominent landform here that emitted dark, fresh-looking pahoehoe to transitional lava flow fields that moved toward the west, but a small arm break into a region where silicic tuff rings dominate a landscape (Fig. 4.54). The scoria cone is composed of fine scoria ash and lapilli that primarily restricted in the volcanic edifice, and just a thin ash plain associated with the volcanic cone. From the scoria cone a perfect view can show the volcanic morphology of an enclosed tuff ring (Tr1), that is a perfect similarity to Jabal Bayda (Fig. 4.54c).

The pyroclastic succession of Tr1 is similar to those recorded in Jabal Bayda, with a potentially more evidence to support involvement of magma and water explosive interaction in the formation of the volcano in the form of abundance of angular, equidimensional low vesicularity (and darker colour) pyroclasts of ash and lapilli, some abundance of crustal-derived accidental lithics and/or xenoliths point toward a potential deeper excavation through the eruptions.

Next to Tr1, another tuff ring provide probably a better access to its proximal to medial pyroclastic succession through a collapsed near vent edifice, exposing about 20 m of pyroclasitc units along the top of the edifice (Fig. 4.55a, b). The pyroclastic succession of Tr2 composed of angular lapilli and block rich fine ash hosted closely packed pyroclastic succession best interpreted to be as a block-and-ash flow deposit. The main pyroclastic units are over 2 m thick and separated by typical cross- and dune-bedded finer grained

Fig. 4.53 Flat floored broad crater of one of the southernmost silicic tuff ring in Harrat Khayber [25° 35′ 14.58″N; 39° 57′ 13.09″E]. The deposits of this tuff ring is simialr to those exposed at Jebel Bayda and dominated with pyzorlcastic density current deposits abundant in fglow banded silicic fragments (**a**). A slightly more eroded tuff ring with rock fall bordered upper rim exposes the tuff ring succession well (circle) allowing to investigate the sediment characterisitcs to infer the volcanic eruption styles dominated the growth of this volcano (**b**). Numbers of "2" and "3" refer to tuff ring 2 and 3

cm-to-dm thick beds (Fig. 4.55b) commonly associated with the formation of block-and-ash flows (Freundt et al. 2000; Ui et al. 1999). The upper part of the succession is more indurated, probably by slight welding and/or some hydrothermal activity that cemented the clasts together to form an erosion resistant capping unit (Fig. 4.55c). From a view point SE of the Tr2 a chain of tuff rings aligned toward Jabal Bayda can be seen that represents a potentially age-defined morphological nature of the tuff rings (Fig. 4.55d).

Just north of Tr2 another slightly better preserved tuff ring is located that has a pyroclastic density current dominated upper succession and a slightly coarser grained lapilli tuff dominated unit in its base. This pyroclastic architecture differs from those recorded from Tr2 and is similar to those identified at Tr1 suggesting a larger variation in eruption style and mechanism in the formation of these volcanoes as we could guess from their general edifice preservation and appearance (Fig. 4.56a).

Interestingly in the eastern side of these chain of tuff rings that form a morphologically characteristic ridge-like massif, a flat floored, low rimmed tuff ring can be seen (Fig. 4.56b). Its crater is filled with aeolian dust suggesting that its original crater floor might have been fairly deep, and the landform might have been a prominent volcanic depression. The great elevation difference between the floor of this and the neighbouring tuff ring crater floors suggest either presence of significant syn-eruptive morphological relief and/or significant age difference between these landform to allow time to cut deeply into the inter-cone relief by fluvial processes and/or different eruption mechanism in the formation of these landforms (e.g. tuff ring versus maar formation). These site by its easy access, relatively small size (500 m across), would make this location a perfect site to apply shallow subsurface geophysical methods to constrain the subsurface architecture of these volcanic edifices. Such future work would also help to constraint the volcanic eruptions scenarios such eruption would pose to the surrounding areas.

In addition to visit the specific silicic tuff rings an effort was taken to locate inter-cone/ring sites where multiple pyroclastic deposits are exposed with an aim to establish the general stratigraphy and the general relationship between variously sourced eruptive products that accumulated in the inter-cone areas. Especially in the middle of the field there were several suitable sites to nominate as key locations to establish the regional volcanic stratigraphy which is a future research goal.

Fig. 4.54 Young scoria cone in the southern region of the Harrat Khayber [25° 35′ 5.55″N; 39° 56′ 16.67″E] emitted transitional type lava flows that enterred in the gap between silicic tuff rings such as Tr1 and Tr2 (**a** and **b**). The Tr1 tuff ring [25° 35′ 14.58″N; 39° 57′ 13.09″E] is nearly perfectly circular with only few gully network and rock-fall dominated erosion scarse

4.4.6 Other Fissure Vents, Lava Shields and Pit Craters

Just north of Jabal Quidr, an extensive lava field and associated spatter and scoria cone dominated volcanic region represent the northern side of Harrat Khaybar. The primary aim to visit this region was to delineate the ash plain associated with Jabal Quidr lateral boundaries. During this exploration work a visit was arranged to one of the main source region of the extensive tube-fed lava field dominate the northern sector of the Harrat Khaybar. One of the most prominent locations is marked as location 4 and it is best described a scoria cone—spatter cone complex with a massive lava tube network (Fig. 4.57).

The main scoria cone has a deep pit crater that is partially mantled by lava drain back features (Fig. 4.57a) recording the fluctuation of lava lake level in the crater. In the inner crater wall abundant, relatively thin lava units suggest a long lived activity persistently released relatively thin lava sheets (Fig. 4.57a). The main crater is connected with a massive lava drain point over 50 m across (Fig. 4.57a) that fed some lava spatter along its margin (Fig. 4.57b) releasing fluidal small-scale (dm to several metres) lava tubes of shelly pahoehoe flows (Fig. 4.57c). In the proximity of this cone other cones form a similar volcanic architecture, commonly feed large tube-fed flows. In steep slopes such tubes commonly disrupted, broken apart, and the lava flow preserved as a rubble pahoehoe mass in a drain channel (Fig. 4.57d).

In the northern side of these vent systems it is evident that the emitted lava flows were dominantly tube-fed, but thin shelly pahoehoe outflows clearly define a lava shield-like architecture (Fig. 4.58a). Main lava tubes are inferred to have accumulated large ponded lava zones where the slope angle has changed (Fig. 4.58b). These ponded lava zones formed fan-like zones that are slightly sink in their middle upon gradual drainage of the flows. In extreme, the roof of these ponded zones collapsed and accumulated a slab-dominated, rubble flow interior confined in circular lava ponds (Fig. 4.58c).

Fig. 4.55 Tr2 tuff ring [25° 36′ 21.69″N; 39° 56′ 5.52″E] pyroclastic succession is exposed in a rock fall cliff face (**a**). The majority of the pyroclastic rocks are angular silicic clast-dominated deposits hosted in variable amount of ash matrix (**b**). The top of the Tr2 pyroclastic succession is moderately welded forming an erosion resistant cap (**c**). From the top of Tr2 a perfect view can be seen toward Tr1 tuff ring [25° 35′ 14.58″N; 39° 57′ 13.09″E] a young scoria cone in the south. Tr2 is a broad tuff ring that is likely to be one of the oldest tuff rings in the region judging from its preservation and the edifice geometry in relationship with the background's topography (**d**)

Fig. 4.56 Panoramic view of the Tr3 tuff ring [25° 38′ 18.99″N; 39° 56′ 10.93″E] crater (**a**). Note the flat floored crater with an initial drainage network and the exposed pyroclastic density current deposits in the inner crater wall. Next to Tr3 a large tuff ring is visible with thick aeolian infill (**b**). In the background other silicic lava domes and the Tr2 tuff ring is visible

4.4.7 Main Findings

The preliminary field survey in Harrat Khaybar confirmed that the field is far easier to access as it was in the past. The field conditions are reasonable good in comparison to Harrat Rahat's conditions, and making this field as an ideal target area for future field work. Geologically the field host numerous silicic tuff rings which are the keys to understand the geochemical evolution of magma from the source conditions through the magma ascent. The explosive nature of the eruptions are evident in many of the silicic tuff rings offering a unique opportunity to understand the nature of dispersed small volume silicic volcanic fields eruption behaviour and develop a realistic eruption scenario-based volcanic hazard study that could be used not only locally but also could serve as a model for understanding similar volcanic fields elsewhere. The young eruptions that formed probably the most complex volcanic edifice in Harrat Khaybar (Jabal Quidr) point to the fact that this volcano likely erupted much longer time than expected from a so called "monogenetic" volcano in intraplate settings and show common features to stratovolcanoes. Its eruption style is likely fall into a sub-Plinian or violent-Strombolian style and therefore its volcanic hazard value for evaluating volcanic eruptions scenarios in western Saudi Arabia, is significant.

4.5 Harrat Al Birk and Tihamat Asir

4.5.1 Overview

In 2014 a field campaign has been arranged to the southern regions of the western Saudi Arabian Cainozoic volcanic fields (Fig. 4.59). The aim of this visit was to assess the field conditions, the general volcanic framework, the accessibility, and the possible geoheritage value of these volcanic fields. In addition the field survey intended to identify key volcanic sites that could be studied in detail to develop a volcanic eruption scenario-based volcanic hazard study on this volcanic region inferred to be active in the past 1 Ma and has some poorly documented historic eruption sites as well (Brown et al. 1989; Coleman and Gregory 1983; Coleman 1993). Thus it can be considered a potentially active volcanic region. The visit was also justified by the fact of several earthquake swarms in the recent past that raised some concern about potential volcanic eruptions in the future. Since the region developing very fast and large cities (Jizan, Sabya, Abu Arish) appear along the coastal area as well as near to the coastal range, the population raise and the infrastructure growth justify well, that such young volcanic field have to be studied in volcanic hazard perspective.

The initial field campaigns were focused on the Al Birk region (Fig. 4.59), mostly along the coast commonly

Fig. 4.57 Scoria cone (C4) and spatter cone (SPC) along a fissure fed an extensive lava flow in the northern part of the Harrat Khayber [25° 47′ 37.23″N; 39° 55′ 57.42″E]. The scoria cone inner crater wall exposes alternating lava flow and scoria beds (**a**). In the norhtern edge of the fissure aligned vent a lava spatter half section exposes the core of a spatter cone (SPC) (**b**) that also emitted low viscosity pahoehoe lava flow fingers forming a cascading lava tube network (**c**). Collapsed lava tubes and confined channelized lava flows are common features in these part of the lava field at Harrat Khaybar (**d**). Fromt eh distance the scoria cone appears to be a prominent volcanic landform (**d**)

Fig. 4.58 Lava shield-like volcanic edifice in the northern sector of the scoria cone [25° 48′ 9.66″N; 39° 55′ 39.89″E]. Note the relatively thin and narrow convolute lava flows that are commonly forming tube network (**a**). In the area where slope angle change and the flow were able to pond, thick crust formed over the lava and large collapsed roofed channels are exposed (**b**). In the wall of such collapsed lava tubes in the wall a series of relatively thin lava layers can be seen (**c**)

referred as Tihamat Asir (*The Coastal Plain of Asir*), a visit to a well-distinguished volcanic area that consists of two individual volcanic complexes (*Jabal Akwa*) relatively far from any other young volcanoes nearby, and a short visit to one of the sites near the coastal range (*Jabal Jirratan*), where some documents record an eruption in the beginning of the last century, however such information have not been confirmed yet.

The Al Birk volcanic field in SW Saudi Arabia is a young intracontinental volcanic field that formed alkaline basaltic volcanoes that are dominated by scoria and spatter cones, extensive lava fields and lava domes/dome coulees (Arno et al. 1980; Brown et al. 1989). The main part [the northern part by Coleman and Gregory (1983)] of the Al Birk volcanic field (Fig. 4.59) spreads over an area 100 km in length and 50 km in width along the Red Sea coast in SW Saudi Arabia (from lat 18° 45′ to lat 17° 45′N) and comprises over 200 individual eruptive centres (Brown et al. 1989; Coleman and Gregory 1983). The main part of the volcanic field is located about 100 km north from the city of Jizan (Fig. 4.59), where a dorsal ridge of older basal lava flows form the base of younger scoria and spatter cones with younger lava flows. Scattered small volcanic regions in the south are separated by the Ad Darb transform fault from the main volcanic region of Harrat Al Birk (Fig. 4.59) are commonly referred to as the harrat of the coastal plain of Asir region, the harrat of Tihamat Asir (Coleman and Gregory 1983). Some work group these volcanic regions as part

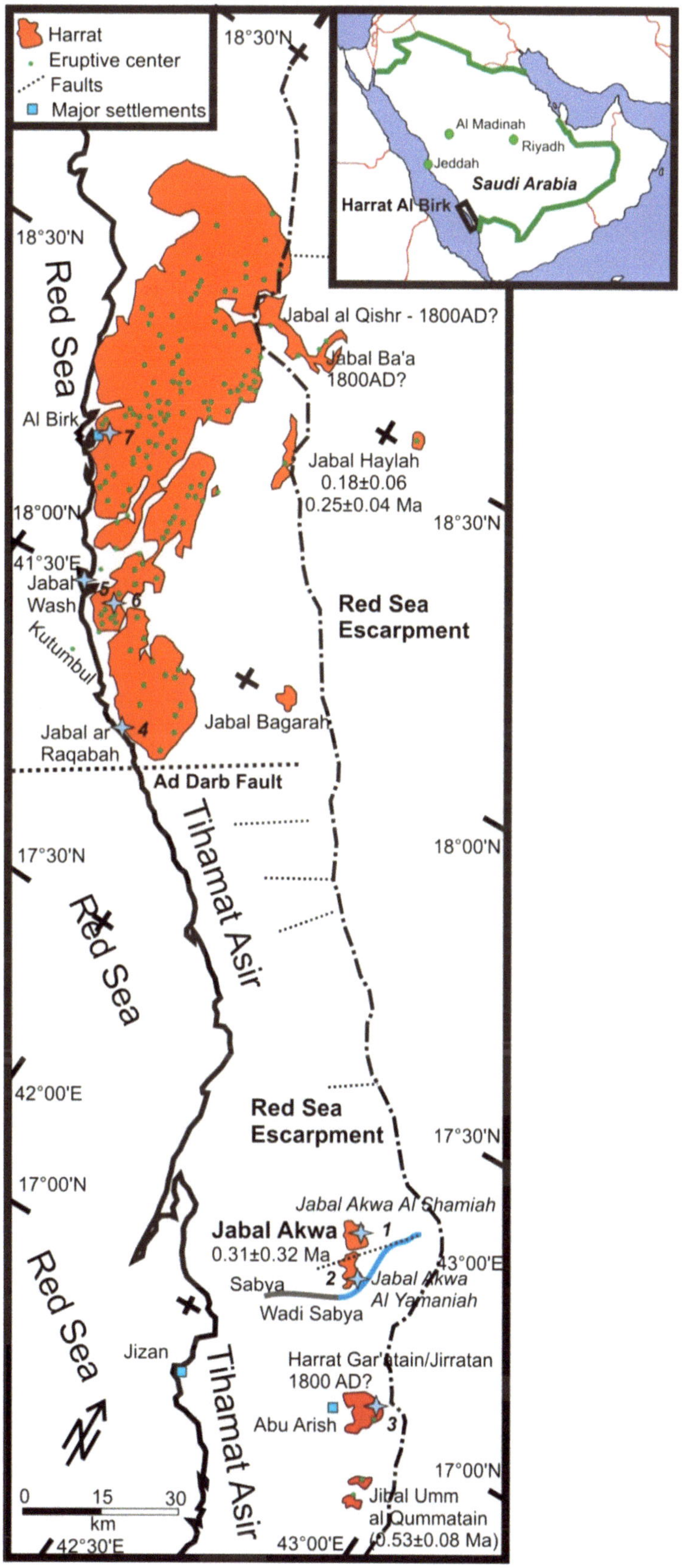

Fig. 4.59 Overview of the Harrat Al Birk. Visited sites are numbered: *1* Jabal Akwa Al Shamiah, *2* Jabal Akwa Al Yamaniah, *3* Harrat Jirratan, *4* Jabal Al Raqabah, *5* Jabal Wash, *6* Khurma, *7* lava fields around Al Birk

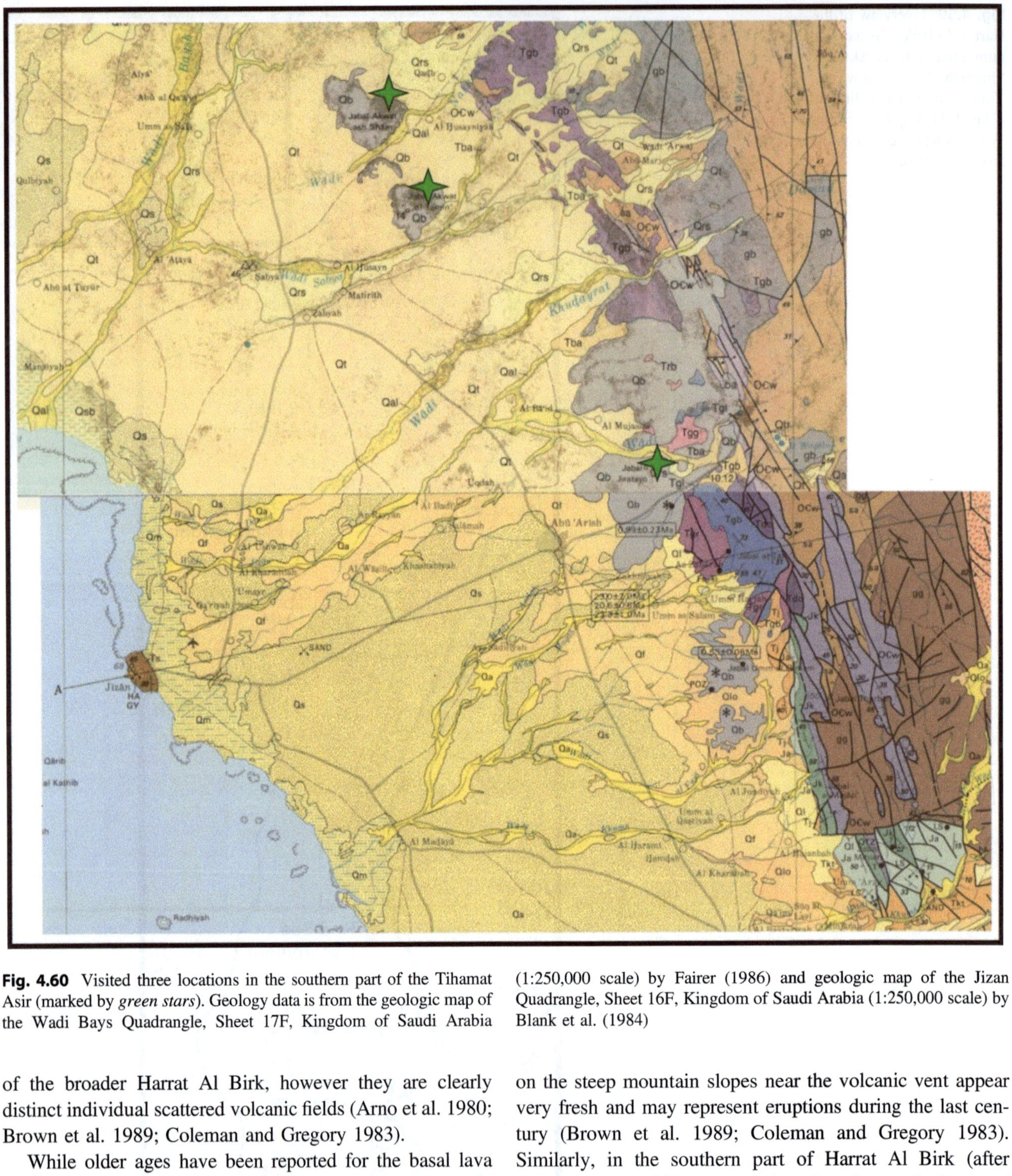

Fig. 4.60 Visited three locations in the southern part of the Tihamat Asir (marked by *green stars*). Geology data is from the geologic map of the Wadi Bays Quadrangle, Sheet 17F, Kingdom of Saudi Arabia (1:250,000 scale) by Fairer (1986) and geologic map of the Jizan Quadrangle, Sheet 16F, Kingdom of Saudi Arabia (1:250,000 scale) by Blank et al. (1984)

of the broader Harrat Al Birk, however they are clearly distinct individual scattered volcanic fields (Arno et al. 1980; Brown et al. 1989; Coleman and Gregory 1983).

While older ages have been reported for the basal lava flows (e.g. as old as 12.4 Ma), some renewed dates have put the age of the volcanic field at less than 2 Ma, with the majority of the eruptions being younger than 1 Ma (Brown et al. 1989; Coleman and Gregory 1983). Some reports have described historic eruptions taking place in the eastern margin of the field, such as Jabal Ba'a and Jabal al Qishr (Fig. 4.59), where isolated patches of black scoriaceous ash on the steep mountain slopes near the volcanic vent appear very fresh and may represent eruptions during the last century (Brown et al. 1989; Coleman and Gregory 1983). Similarly, in the southern part of Harrat Al Birk (after Coleman et al. 1983), commonly referred to as the harrat of Tihamat Asir—the coastal plain of Asir (Vincent 2008), an eruption in the last century was reported in an area called the Harrat Gar'atain (Jiratan) (Fig. 4.59) (Neumann Van Padang 1963). These reported (but only loosely confirmed) young volcanic eruptions of the Harrat Al Birk and the Harrat of Tihamat Asir are associated with volcanic cones that appear

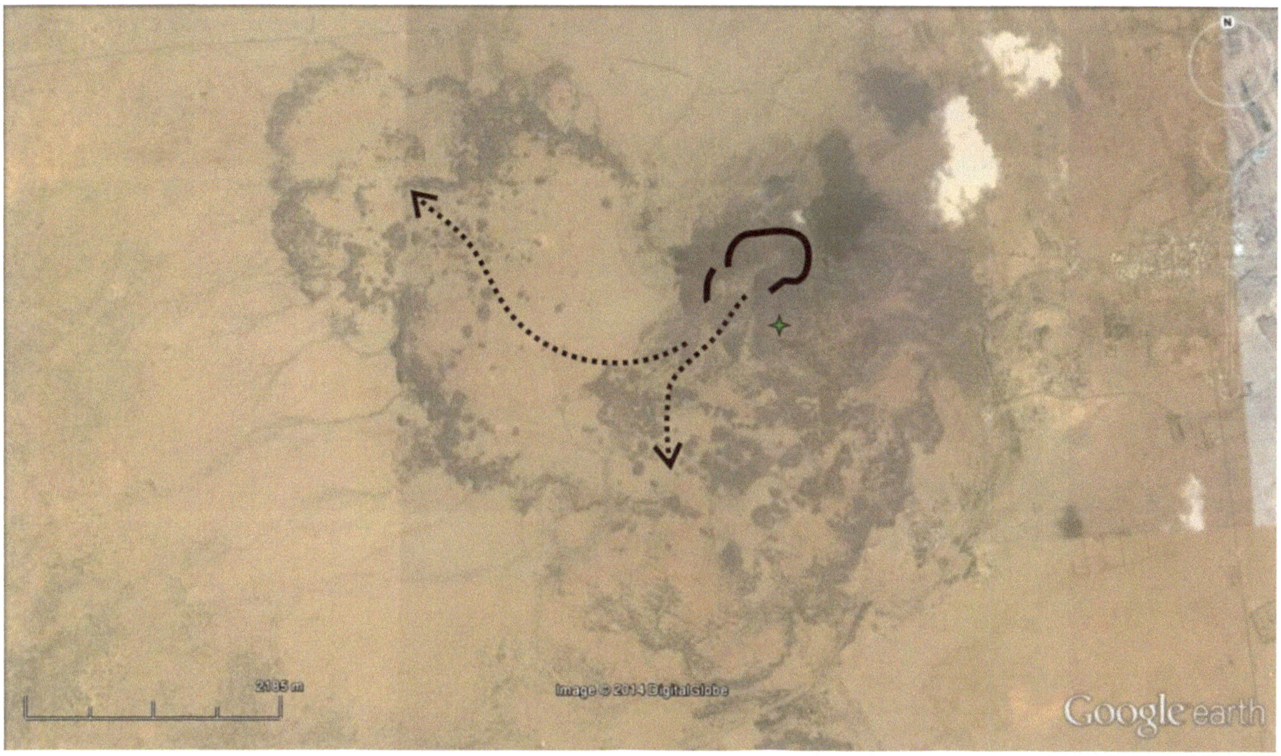

Fig. 4.61 Google earth image of Jabal Akwa Al Shamia [17° 15′ 13.99″N; 42° 42′ 59.66″E]. Continuous line represents the crest of the preserved crater rims. *Dashed arrows* show lava outflows feeding large lava fans in the western side of the cones. *Green star* mark the top of the quarry where the pyroclastic succession of the scoria cones can be observed

to be very youthful in their morphology, vegetation cover, and erosion level, indicating that this region in SW Saudi Arabia is a potentially active volcanic area, and therefore detailed study of these young volcanoes is essential to shed light on potential future volcanic eruption scenarios with which the region may be faced. The general view of the volcanoes formed in the past 1 Ma in Harrat Al Birk and the Harrat of Tihamat Asir is that the harrat is largely dominated by scoria and spatter cones associated with extensive lava flows. Identification of any evidence of phreatomagmatic explosive eruptions in the recent history of this volcanic field is critical in terms of defining volcanic hazards that are considered to be fast, destructive and highly unpredictable.

4.5.2 Jabal Akwa Al Shamiah as a Complex Volcanic Geotope [17° 15′ 13.99″N; 42° 42′ 59.66″E]

South of the main body of Harrat Al Birk, just NE from Jizan city, near the city of Sabya (Fig. 4.59), two well-distinguished large scoria cones and associated lava fields dominate the landscape: Jabal Akwa Al Shamiah in the north and Jabal Akwa Al Yamaniah in the south (Fig. 4.60). Both volcanoes (called Akwa cones in English, or Jibal Akwa in Arabic) are composed of a large (about 100 m high) scoria cones with a breached crater toward the west, an extensive lava field and some mounds that are inferred to be rafted cone pieces or distal lava flow fronts, which stalled and formed piles of lava about 3 km from their sources (Fig. 4.61). Both cones are covered by aeolian deposits that make the identification of their volcanic facies difficult (Fig. 4.61).

The northern cone complex, Jabal Akwa Al Shamiah has a complex morphology with at least two well-distinguished craters (Fig. 4.61), each breached toward the west and composed of steeply dipping lava spatter and spindle-bomb rich proximal edifice-building pyroclastic units (Fig. 4.62a, b). The inner crater wall has been preserved in a few places and is defined by agglutinated spatter that forms an erosion-resistant collar on the lip of the crater, indicating that, at least in the final stage of the volcanic activity, this cone had lava fountain dominated eruptions (Fig. 4.62a). In contrast, the main edifice is composed of black and red scoria lapilli and ash, which forms a well-developed cone edifice that is gradually transforming into an inter-cone ash plain traceable over 3 km from the cone (Fig. 4.62c). The scoriaceous ash and lapilli is angular, moderately to highly vesicular and the clasts are primarily isometric, but more flattened clasts are also known, indicating fluctuation between normal Strombolian style

Fig. 4.62 *Panoramic views* to Jabal Akwa Al Shamia [17° 15′ 13.99″ N; 42° 42′ 59.66″E]. The main cone is complex scoria cone (**a**) that is breached toward the SW from where a lava field fed (JAY = Jabal Akwa Al Yamaniah). In the medial to a distal part of the lava flow some spatter-rich mound field can be observed (**a** and **b**). The pyroclastic succession of the main cone (**c**) composed of black to grey vesicular equidimensional scoria ash and lapilli. Scoriaceous tuff breccias (*stb*) are commonly host large lava spatters and fusiform bombs. The *upper* part of the section is dominated by a scoria-rich black ash (*sa*) and grey matrix rich lapilli tuff (*mlt*)

eruptions and lava fountaining (Fig. 4.62c). Large chunks of former cone flanks form a mound zone on a lava field spread isometrically toward the western side of the cone, suggesting that the cone erupted in a gentle westward dipping coastal plain. In the south, another cone complex is known (Fig. 4.62a, b).

4.5.3 Jabal Akwa Al Yamaniah as a Complex Volcanic Geotope [17° 11′ 26.18″N; 42° 44′ 6.48″E]

Jabal Akwa Al Yamaniah is a similar size to Jabal Akwa Al Shamiah (Fig. 4.60); however, this cone is more eroded in its northern flank, forming a steep and evenly spaced gully network (Fig. 4.63a). The cone is surrounded by two circular lava flows. An upper lava field, which has a flow front in the west, reaches about 3 km from the cone, while the other lava field spread about 6 km from the cone and formed a lava surface about 50 m below the other lava field. The cone has a large breach toward the west, and the upper lava flow front and the cone area is covered by a thick aeolian Quaternary succession. In addition, the gently westward-dipping, Quaternary sedimentary cover hosts numerous archaeological sites with pottery remains and grinding stones, suggesting that this volcano has hosted a village in the past (Fig. 4.63a).

Undescribed tuff deposits have been noted below the Pleistocene lava flows intercalated with fluvial terrace and aeolian deposits in the south of Jabal Akwa Al Yamaniah, along the Wadi Sabya; however, their origin was not known

Fig. 4.63 The main scoria cone of Jabal Akwa Al Yamaniah [17° 11′ 26.18″N; 42° 44′ 6.48″E] (**a**) is breached toward the west and it has a well-developed gully network on its outer flank. The top of the volcanic massif is covered by aeolian deposits on what archaeological sites reveal a presence of a large human settlement that were active during wet climatic periods. The southern and western side of the volcanic massif (**b**) exposes a tuff ring (*tr*) rim, that is partially overrun by a lava flow that were inferred to infill a large volcanic crater a maar. The debris fan (tr-df) that covers the tuff ring made it difficult to identify the tuff ring itself. The boundary between the crater-filling lava—overspill lava flow (*cflf*) is marked by *yellow dashed line* on (**b**). The tuff ring is dominated by accidental lithic rich lapilli tuff and tuff deposited from base surge dominated currents (**c** and **d**). In the southern edge of the volcanic massif the tuff ring is partially covered by aeolian dust (*as*) and backed the lava flow (*lf*) confining it in the crater. Transportation indicators (*white arrow* on **c** and **d**) show base surge transportation from the volcanic massif

(Dabbagh et al. 1984). An accidental lithic rich tuff and lapilli tuff succession has been identified in the southern and eastern margin of the thick upper lava flow fronts of Jabal Akwa Al Yamaniah (Fig. 4.63b). The tuff and lapilli tuff deposits dip about 15–20° away from the cone, forming the well-defined remains of a former gently sloping volcanic edifice in these sectors of the volcano (Fig. 4.63c, d). The pyroclastic succession is composed of angular glassy pyroclasts (lapilli and ash sized), with a majority being partially or completely altered, red to brown palagonite, abundant various sizes of accidental lithics of known crustal rock types, and mantle-derived nodules (Fig. 4.63c, d). The pyroclastic succession shows a general trend of fining upward, having a lapilli tuff succession that is more lithic-dominated at its base, which gradually transforms to a better sorted, finer grained and more juvenile pyroclast-rich, coarse-fine succession at the top of the section (Fig. 4.63c, d). The accidental lithic fragments are commonly well-rounded gravels, while silts and mud-stones are common among the large angular lithics. The upper part of the succession is more dune and cross-bedded with abundant features recording the deposition from horizontal moving pyroclastic density currents and the microtopography of the depositional surface (Fig. 4.63c, d). The total thickness of this pyroclastic succession is about 40 m in the SE.

The abundant glassy to palagonitized, low vesicularity juvenile pyroclasts indicate fast chilling of the fragmented magma, which is consistent with magma-water interaction triggered explosive fragmentation (White and Ross 2011). The abundance of a great variety of country rocks as accidental lithics in the pyroclastic units indicates that the explosions excavated a significant proportion of the bed rocks, suggesting that the magma and water interaction took place below the syn-eruptive surface, such as in the case of a maar-forming volcanic eruption (White and Ross 2011). The dune-bedded and unsorted nature of the majority of the pyroclastic succession is consistent with an origin from a base surge dominated eruption (White and Ross 2011).

The at least 40 m thick pyroclastic succession is inferred to be part of a former tuff ring that is today partially engulfed by post-eruptive, Quaternary aeolian and fluvial terrace deposits. The tuff ring formed a barrier to the intra-crater lava flows emitted from the spatter cones that were formed in the crater of the initial phreatomagmatic volcano. The accidental lithic-dominated pyroclastic deposits suggest that the explosive eruptions must have excavated a significant portion of country rocks and formed a maar volcano that is surrounded by its tuff ring in the ancestral Wadi Sabya. A crater was carved in the wadi deposit and subsequently functioned as a depocentre, collecting lava flows emitted from the scoria and spatter cones in the maar crater. By the complete exhaustion of the available ground-water supply, the initial phreatomagmatic explosive eruptions led to the formation of a scoria cone similar to common trends associated with monogenetic volcanoes elsewhere (Kereszturi and Németh 2012), probably on the syn-eruptive surface outside of the maar, on its northern side. The large size and complex stratigraphy of the scoria cone indicates that it was erupted over a prolonged period of time.

Jabal Akwa Al Yamaniah is the first volcano where a phreatomagmatic pyroclastic succession has been identified in SW Saudi Arabia in the Tihamat Asir. It records a complex volcanic eruptive history that started with a violent explosive eruptive phase, when rising magma interacted with ground-water and formed a deeply excavated crater: a maar surrounded by a tuff ring. Mapping indicates that the tuff ring was probably thicker in the SE and absent in the northern sector of the volcano. The circular distribution of the lava flows and spatter mounds of Jabal Akwa Al Yamaniah is the direct result of the lava flows being controlled by the crater rim of the initial phreatomagmatic volcano, capturing and collecting post-maar eruptive products in a broad maar crater. The identification of the basal maar volcano in this location reveals for the first time that the explosive interaction between magma and water needs to be evaluated seriously in this region as a potential high consequence volcanic hazard similarly to other coastal regions elsewhere (Agustin-Flores et al. 2014; Brand et al. 2014; Németh et al. 2012a). Alluvial fans, wadi deposits or deep faults, especially along the coastal escarpment, can host significant volumes of water (particularly in the rainy season) and this is capable of dramatically changing the volcanic eruption style of the rising magma, making it more destructive and hazardous.

A complex of phreatomagmatic volcanoes has been identified for the first time in the southernmost portion of the Al Birk volcanic field in SW Saudi Arabia. The newly identified accidental lithic clast-rich tuff and lapilli tuff succession is partially covered by aeolian sand and wadi deposits, with abundant mantle- and deep crustal-derived xenoliths in the southern margin. This pyroclastic succession of the Jabal Akwa Al Yamaniah volcanic cone complex is typical of a volcano that had phreatomagmatic explosive eruptions in its initial eruptive stage. The large volume of accidental lithics in this basal pyroclastic succession indicates that this volcano is a maar-diatreme and its eruption was triggered by the explosive interaction of rising magma and ground water in a thick gravelly alluvial plain cross-cut by wadi networks. The young age (<1 Ma) of the Al Birk volcanic field in general puts this discovery in the spotlight, as it provides firm evidence that phreatomagmatism cannot be neglected, at least in the initial stage of any future eruptions—especially those that occur over thick alluvial fans in the coastal regions of Jizan—and should be viewed as a potentially destructive and highly unpredictable, high impact volcanic hazard.

Fig. 4.64 Youthful appearance of the Jirratan scoria cone [17° 0′ 44.01″N; 42° 54′ 12.45″E] that was reported to be active in the beginning of the last century. *Arrow points* to another scoria cone that could also be the mystery young scoria cone in Harrat Jirratan

4.5.4 Harrat Jirratan [17° 0′ 44.01″N; 42° 54′ 12.45″E]

A short field visit explored the vicinity of Harrat Jirratan, a region mentioned as hosting a volcano that erupted about 100 years ago (Neumann Van Padang 1963). This initial visit aimed to assess the site and to provide some recommendation for research strategy to understand the geological context of the region. Harrat Jirratan is located near to the town of Abu Arish, along the main highway from Jizan to Yemen (Fig. 4.64). The location is the border zone between KSA and Yemen and therefore it is difficult to access sites. One of the suspected sites for instance is a military base where access was not permitted (Fig. 4.64). From the distance however the cone look fresh and indeed could be as young as a century. In the other hand another cone (Fig. 4.64) that were more likely the mentioned historic site can be accessed but the cone itself is surrounded by fence, and direct access to the main part of the cone is not allowed without appropriate administrative preparations. The area however consists of a lava field of blocky lava forming a characteristic plateau (Fig. 4.64) above the town of Abu Arish. The lava field surface morphology is difficult to assess due to the exposed gently westward dipping slopes that are brushed by strong winds and therefore aeolian dust pressed into the gaps of any lava surface (Fig. 4.64). The lava flows however seem to be thin micropahoehoe styles and indeed they share some evidence that they cannot be too old. The cones are also steep, their crater rim is well-preserved and the flank of the cones has no gully network visible. These morphological features point to a relatively young age of the cones however not providing enough evidence to support their very young age. The difficulty is in this area that the elevation of the cones, and the occasional rainfall able to sustain some scrub and grass vegetation that makes the landscape fairly similar to those known from Chyulu Volcanic Field in eastern Kenya (Shaitani and Chainu eruptions in 1856) [http://www.volcano.si.edu/volcano.cfm?vn=222130], where recent scoria cones look also older than they are (Haug and Strecker 1995).

The scoria cones are composed of red vesicular scoria that is isometric in shape. Fine black ash is occasionally visible in the upper section of the edifice. The cones have an enclosed crater that is not filled with aeolian dust indicating its relatively young age.

4.5.5 Jabal Al Raquabah [17° 48′ 40.12″N; 41° 50′ 35.49″E]

Jabal Al Raquabah is an eroded volcanic edifice located directly on the Red Sea coast in the southern part of the main body of the Harrat Al Birk. The erosional remnant can be accessed all around and there are some tracks that cross cut the edifice allowing a fairly complex view for the visitor. The edifice is not covered by any vegetation however its north facing site has a significant aeolian sand dune cover about one third up on its flank. The southern side of the edifice however perfectly exposed and allow a complete

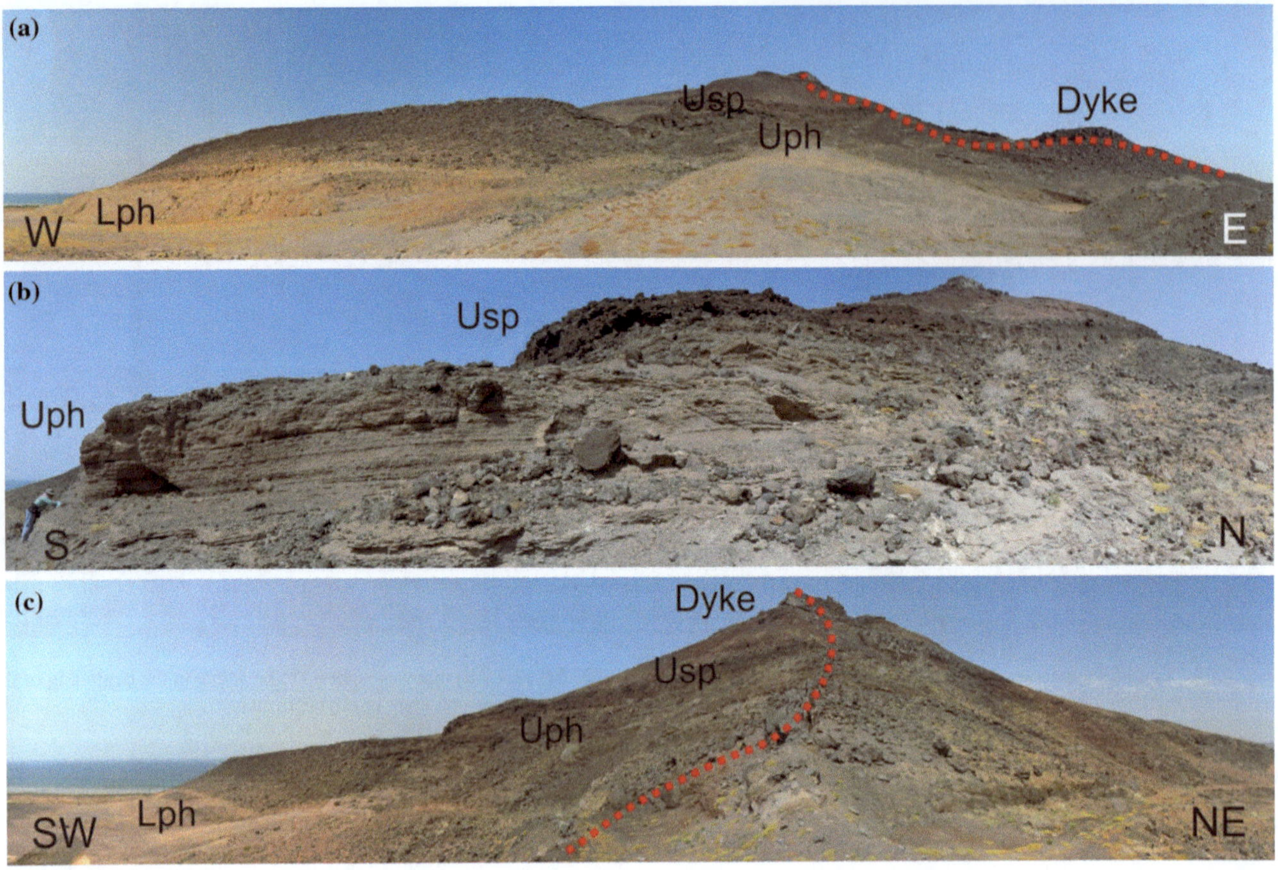

Fig. 4.65 Overview images of Jebel Raquabah volcano [17° 48′ 40.12″N; 41° 50′ 35.49″E] that is inferred to be an erosional remnant of a Surtseyan emergent volcano. Its base composed of phreatomagmatic deposits (*Lph*) deposited from pyroclastic density currents that is gradually transforming into a more fall-dominated well-bedded, glassy pyroclast-rich succession (*Uph*) (**b**). The entire succession is capped by a spatter dominated partially agglutinated unit (*Usp*) that is framed with some dykes (**c**)

section to be measured from the sea level up to the top of the cone. The volcanic edifice is a conical shape broad volcanic massif about 3 km across with a small crater on top of the entire basal succession (Fig. 4.65a). The base of the pyroclastic succession forms an about 15 m thick succession of glassy pyroclast rich bedded lapilli tuff and tuff that is rich in ballistic bombs of angular chilled juvenile lapilli and bomb (Fig. 4.65a). The abundance of glassy pyroclasts, the dune to cross bedded nature of the deposits and presence of accidental lithic clasts as accidental lithics suggest that these beds formed through phreatomagmatic explosive eruption. The relative low proportion of deep seated accidental lithics, the abundance of glassy pyroclasts and presence of coral and other shallow marine sediment-derived clasts suggest that the eruptions were triggered by interaction of rising magma and shallow seawater behind a reef and the eruption were likely initiated in a shallow subaqueous setting forming an emergent, Surtseyan style volcano (White and Houghton 2000). The water depth could have been very low and at this stage it is difficult to say that the volcano evolved as a typical Surtseyan volcano (Agustin-Flores et al. 2015; Brand and Clarke 2009; Mattsson 2010; Sohn et al. 2008), or it was just strongly influenced by the availability of surface water and otherwise evolved as a tuff ring on a marshy coastal plain (Agustin-Flores et al. 2014) or phreatomagmatic rift-edge volcanoes commonly develop on volcanic islands coastal regions (Németh and Cronin 2009, 2011).

The stratigraphy of Jabal Raquabah is simple by having a basal phreatomagmatic succession dominated by pyroclastic density current deposited beds (Lph) that gradually transform to an upper phreatomagmatic succession (Uph) that is more clast-supported and more consistent with fall deposition in sub aerial conditions during the fully emergent stage of the volcano, and an upper spatter dominated part (Usp) that is a capping unit and clearly demonstrate the complete emergent and/or water cut of stage of the eruption (Fig. 4.65).

Texturally the basal successions are rich in horizontal transport indicators (Fig. 4.66a, b). About 15 m above the base of this initital succession the pyroclastic beds have a zone of about 2 m where multiple flow indicators

Fig. 4.66 **a** Basal pyroclastic succession of Jabal Al Raquabah [17° 48′ 40.12″N; 41° 50′ 35.49″E], **b** matrix supported part of the basal phreatomagmatic succession of Jabal Al Raquabah, **c** transitional multiple transport indicator-dominated section of Jabal Al Raquabah, **d** coral fragments in the phreatomagmatic successions, **e** clast supported lapilli bed in the upper phreatomagmatic unit with great variety of palagonitization (various brown coloured clasts) and some coral fragments (*white clasts*), **f** upper lava spatter dominated capping unit

demonstrate a multioriented transportation regime best explained by a wave action dominated transportation region in shallow coastal regions (Fig. 4.66c). The basal section and the upper phreatomagmatic units contain large number of coral fragments (Fig. 4.66d) as accidental clasts indicating that these clasts were excavated by an explosive eruption. The upper phreatomagmatic succession clast supported beds show variable palagonitization of individual pyroclasts suggesting an intensive recycling in a water, steam and acid-rich environment (Fig. 4.66e). The capping succession is typical lava spatter dominated unit that can be more matrix supported in what large spatter float or more

Fig. 4.67 Khurma scoria cone [18° 13′ 32.99″N; 41° 36′ 29.77″E] sits on an elevated plateau (**a**). The cone itself is dissected, partially collapsed, and its edifice interior is exposed (**b**)

Fig. 4.68 Al Birk town [18° 12′ 47.56″N; 41° 32′ 17.70″E] from the lava plug forming a hill in the centre of the town (**a**). The centre of the town form an elevated coherent lava body that is surrounded by a "wall" of circular lava flows (*arrows* on **b**). The origin of these features is not clear, and could be a subject of future research

homogeneous up-section an form agglutinated and welded beds that acted as an erosion resistant cap on the volcanic edifice (Fig. 4.66f).

4.5.6 Khurma [18° 13′ 32.99″N; 41° 36′ 29.77″E]

Khurma is a large scoria cone dominates the central part of the Harrat Al Birk (Fig. 4.67). It sits over a blocky lava flow field that considered to be an older paleosurface. The cone is complex and shows multiple collapse and rebuild events in their exposed sections (Fig. 4.67b). It has no associated ash plain around the cone, however it could have been eroded since the volcano sits over a wind exposed plateau (Fig. 4.67a). The location is accessible and could be an important easy to access and nice geosites for a future geopark in the region.

4.5.7 Al Birk [18° 12′ 47.56″N; 41° 32′ 17.70″E]

The town of Al Birk is the location after the Harrat Al Birk got its name. The city center seems to sit on an elevated platform that stands out from a depression that is surrounded by a circular feature. This geometrical set was the reason the field visit was arranged. Such setting is commonly linked with a maar-diatreme feature. However, the circular feature turned to be a lava flow front that is by some reason stopped in a semi-circular fashion. In the center of this circulate feature a large cliff with alkaline basaltic rocks form a prominent landmark (Fig. 4.68). Toward the Red Sea similar older looking alkaline basaltic lava flows commonly form a paleo-surface against old reef banked suggesting some sea level rise and fall since these lava fields formed. The age and chemical composition of these lavas could provide interesting information on the general landscape forming lava flow fields in the core of the Harrat Al Birk.

4.5.8 Main Findings

Harrat Al Birk is an interesting volcanic field in term of the variety of volcanic landforms identified in this field campaign. First time a partially buried maar volcano have been located at Jabal Akwa, that highlight the need for further detailed studies in the region and evaluate the volcanic hazard aspects of this finding in regard of linking the location for these vents and the location of major wadi networks. Also, along the Red Sea coast the first time phreatomagmatism were noticed. The described features point to infer a shallow submarine, emergent tuff cone growth that fits very well for the expected eruption scenarios during higher sea level one can envision. The centre part of the Harrat Al Birk is also promising in regard of the large partially eroded scoria cones that allow looking inside their edifice and characterises their volcanic eruptions. Interestingly, there are several cones that are fairly steep, and suggest that some sort of lava dome activity may have played a role in the evolution of these cones.

Harrat Al Birk is located in a fast growing region where population growth and investment increase is huge. The Harrat Al Birk is a potentially active volcanic field, and therefore it is important to fully develop eruption scenario hazard models and communicate those scientific results with local communities and authorities. The newly identified volcanic eruption styles in the Harrat Al Birk made it important that such eruption scenario studies conducted in the near future.

References

Abdel Wahab A, Abul Maaty MA, Stuart FM, Awad H, Kafafy A (2014) The geology and geochronology of Al Wahbah maar crater. Quaternary Geochronology, Harrat Kishb, Saudi Arabia 21:70–76 doi:10.1016/j.quageo.2013.01.008

Agustin-Flores J, Nemeth K, Cronin SJ, Lindsay JM, Kereszturi G, Brand BD, Smith IEM (2014) Phreatomagmatic eruptions through unconsolidated coastal plain sequences, Maungataketake, Auckland Volcanic Field (New Zealand). J Volcanol Geoth Res 276:46–63

Agustin-Flores J, Nemeth K, Cronin SJ, Lindsay, JM, Kereszturi G (2015) Construction of the north head (Maungauika) tuff cone: a product of Surtseyan volcanism, rare in the Auckland Volcanic Field, New Zealand. Bull Volcanol 77(2)

Al-Harthi AA (1998) Effect of planar structures on the anisotropy of Ranyah sandstone, Saudi Arabia. Eng Geol 50(1–2):49–57

Al-Rehaili M, Shouman S (1985) Ground fracture and damage to buildings at an Na'ay village (27/42B), Kingdom of Saudi Arabia. Saudi Arabian Directorate General of Mineral Resources Open-File report DGMR-OF-06-1

Anderson SW, Smrekar SE, Stofan ER (2012) Tumulus development on lava flows: insights from observations of active tumuli and analysis of formation models. Bull Volc 74(4):931–946

Arno V, Bakashwin MA, Baker AY, Barberi F, Basahel A, Dipaola GM, Ferrara G, Gazzaz MA, Giuliani A, Heikel M, Marinelli G, Nassief AO, Rosi M, Santacroce R (1980) Recent basic volcanism along the Red Sea coast: the AI Birk lava field in Saudi Arabia. In: Zanettin B (ed) Geodynamic evolution of the Afro-Arabian rift system. Accademia Nazionale dei Lincei, Rome, pp 645–654

Austin-Erickson A (2007) Phreatomagmatic eruptions of rhyolitic magma; a case study of Tepexitl tuff ring Serdan-Oriental Basin, Mexico [Unpublished Master's Thesis]. In: Northern Arizona University, Flagstaff, pp 234–234

Austin-Erickson A, Buettner R, Dellino P, Ort MH, Zimanowski B (2008) Phreatomagmatic explosions of rhyolitic magma; experimental and field evidence. J Geophys Res 113(B11):0–Citation B11201

Austin-Erickson A, Ort MH, Carrasco-Nunez G (2011) Rhyolitic phreatomagmatism explored: Tepexitl tuff ring (Eastern Mexican Volcanic Belt). J Volcanol Geoth Res 201(1–4):325–341

Baker PE, Brosset R, Gass IG, Neary CR (1973) Jebel al Abyad: a recent alkalic volcanic complex in western Saudi Arabia. Lithos 6:291–314

Bankher KA, Al-Harthi AA (1999) Earth fissuring and land subsidence in western Saudi Arabia. Nat Hazards 20(1):21–42

Blaikie TN, Ailleres L, Cas RAF, Betts PG (2012) Three-dimensional potential field modelling of a multi-vent maar-diatreme; the Lake Coragulac maar, Newer Volcanics Province, south-eastern Australia. J Volcanol Geoth Res 235–236:70–83

Blaikie TN, Ailleres L, Betts PG, Cas RAF (2014) Interpreting subsurface volcanic structures using geologically constrained 3-D gravity inversions: examples of maar-diatremes, Newer Volcanics Province, southeastern Australia. J Geophys Res Solid Earth 119 (4):3857–3878

Blank HR, Johnson PR, Gettings ME, and Simmons GC (1984) Explanatory notes to the geologic map of the Jizan quadrangle, sheet 16F, Kingdom of Saudi Arabia: Saudi Arabian Deputy Ministry of Mineral Resources Geoscience Map GM-104C

Blusztajn J, Hart SR, Shimizu N, McGuire AV (1995) Trace-Element and Isotopic Characteristics of Spinel Peridotite Xenoliths from Saudi-Arabia. Chem Geol 123(1–4):53–65

Bogaard PVD, Schmincke H-U (1984) The eruptive center of the late quaternary Laacher see Tephra. Geol Rundsch 73(3):933–980

Boyce J (2013) The newer volcanics province of southeastern Australia: a new classification scheme and distribution map for eruption centres. Aust J Earth Sci 60(4):449–462

Brand BD, Clarke AB (2009) The architecture, eruptive history, and evolution of the table rock complex, Oregon: from a Surtseyan to an energetic maar eruption. J Volcanol Geoth Res 180(2–4):203–224

Brand BD, Gravley DM, Clarke AB, Lindsay JM, Bloomberg SH, Agustin-Flores J, Németh K (2014) A combined field and numerical approach to understanding dilute pyroclastic density current dynamics and hazard potential: Auckland Volcanic Field, New Zealand. J Volcanol Geoth Res 276:215–232

Brown GF, Schmidt DL, Huffan Jr AC (1989) Geology of the Arabian Peninsula, Shield area of western Saudi Arabia,. U.S. Geological Survey professional paper, 560-A:1–188

Calvari S, Pinkerton H (1999) Lava tube morphology on Etna and evidence for lava flow emplacement mechanisms. J Volcanol Geoth Res 90(3–4):263–280

Camp VE, Roobol MJ, Hooper PR (1991) The Arabian continental alkali basalt province 2. Evolution of Harrats Khaybar, Ithnayn, and Kura, Kingdom of Saudi-Arabia. Geol Soc Am Bull 103(3):363–391

Camp VE, Roobol MJ, Hooper PR (1992) The Arabian continental alkali basalt province 3. Evolution of Harrat Kishb, Kingdom of Saudi-Arabia. Geol Soc Am Bull 104(4):379–396

Campos Venuti M, Rossi PL (1996) Depositional facies in the Fyriplaka rhyolitic tuff ring, Milos Island (Cyclades, Greece). Acta Vulcanol 8(2):173–190

Cano-Cruz M, Carrasco-Nunez G (2008) Evolution of a rhyolitic explosion crater (maar): Hoya de Estrada, Valle de Santiago volcanic field Guanajuato. Mexico. Revista Mexicana De Ciencias Geologicas 25(3):549–564

Carter A, Wyk de Vries B, Kelfoun K, Bachelery P, Briole P (2007) Pits, rifts and slumps: the summit structure of Piton de la Fournaise. Bull Volc 69(7):741–756

Chagarlamudi P, Zakir FA, Moufti MR, Rogers RH (1991) Application of aerial phtography, Landsat TM and radar images in delineating volcanic features in Harrat Khaybar, Kingdom of Saudi Arabia. Proc Thematic Conf Geol Remote Sens 8:613–626

Coleman RG (1993) Geologic evolution of the Red Sea. Oxford Monographs on Geology and Geophysics, 24 [Oxford University Press, New York]:1–186

Coleman RC, Gregory R (1983) Cenozoic volcanic rocks. U. S. Geological Survey Professional Paper 287–287

Coleman RG, Gregory RT, Brown GF (1983) Cenozoic volcanic rocks of Saudi Arabia. Saudi Arabian. Deputy Minister of Mineral Resources, Open File Report 590 USGS-OF93:1–82

Connor CB, Conway FM (2000) Basaltic volcanic fields. In: Sigurdsson H (ed) Encyclopedia of Volcanoes. Academic Press, San Diego, pp 331–343

Cronin S, Németh K, Smith I, Leonard G, Shane P (2009) Possible rejuvenation of volcanism at the "monogenetic" phreatomagmatic/magmatic volcanic complex of Panmure Basin, Auckland Volcanic Field, New Zealand. In: Haller MJ, Massaferro GI (eds) Abstract Volume of the IAVCEI-IAS 3rd International Maar Conference (Malargue, Argentina) 23–24

Dabbagh A, Emmermann R, Hoetzl H, Jado AR, Lippolt HJ, Kollmann W, Moser H, Rauert W, Zoetl JG (1984) The development of Tihamat Asir during the Quaternary. In: Jado AR, Zoetl JG (eds) The Quaternary Period in Saudi Arabia, vol 2., Springer-VerlagVienna, Austria, pp 150–173

Demange J, Baubron J-C, Marcelot G, Cotten J, Maury RC, Anonymous (1983) Cadre structural, petrologie et geochimie de la serie volcanique de Jabal al Abyad (Arabie Saoudite). Bulletin des Centres de Recherches Exploration-Production Elf-Aquitaine 7 (1):232–248

Druitt TH, Brenchley PJ, Gokten YE, Francaviglia V (1995) Late Quaternary rhyolitic eruptions from the Acigol Complex, central Turkey. J Geol Soc London 152(4):655–667

Duncan RA, Kent AJR, Thornber CR, Schlieder TD, Al-Amri AM (2016) Timing and composition of continental volcanism at Harrat Hutaymah, western Saudi Arabia. J Volcanol Geotherm Res 313: 1–14

Duraiswami RA, Bondre NR, Dole G (2004) Possible lava tube system in a hummocky lava flow at Daund, western Deccan Volcanic Province, India. Proc Indian Acad Sci Earth Planet Sci 113(4):819–829

Fairer GM (1986) Geologic map of the Harrat Ithnayn Quadrangle, sheet 26D, Kingdom of Saudi Arabia. In: Directorate General of Mineral Resources, Ministry of Petroleum and Mineral Resources, Jiddah, Saudi Arabia (SAU), pp 15

Fisher RV, Schmincke H-U, van den Bogaard P (1983) Origin and emplacement of a pyroclastic flow and surge unit at Laacher See, Germany. J Volcanol Geoth Res 17:375–392

Freundt A, Wilson CJN, Carey SN (2000) Ignimbrites and block-and-ash flow deposits. In: Sigurdsson H, Houghton BF, McNutt SR, Rymer H, Stix J (eds) Encyclopedia of Volcanoes. Academic Press, San Diego, pp 581–600

Gondal MA, Nasr MM, Ahmed Z, Yamani ZH (2009) Determination of trace elements in volcanic rock samples collected from cenozoic lava eruption sites using LIBS. J Environ Sci Health A 44(5):528–535

Grainger DJ (1996) Al Wahbah volcanic explosion crater, Saudi Arabia, Geol Today, Jan–Feb:27–30

Grainger DJ, Hanif MR (1989) Geologic map of the Shaghab Quadrangle, Sheet 27B, Kingdom of Saudi Arabia. Directorate General of Mineral Resources, Ministry of Petroleum and Mineral Resources, Jiddah, Jiddah, Saudi Arabia (SAU), pp 0–1 sheet

Guilbaud MN, Blake S, Thordarson T, Self S (2007) Role of syn-eruptive cooling and degassing on textures of lavas from the AD 1783-1784 Laki eruption, south Iceland. J Petrol 48(7):1265–1294

Gutmann JT (1976) Geology of Crater Elegante, Sonora, Mexico. Geol Soc Am Bull 87:1718–1729

Gutmann JT (2002) Strombolian and effusive activity as precursors to phreatomagmatism; eruptive sequence at maars of the Pinacate volcanic field, Sonora, Mexico. J Volcanol Geoth Res 113(1–2):345–356

Harris AJL, Favalli M, Mazzarini F, Hamilton CW (2009) Construction dynamics of a lava channel. Bull Volc 71(4):459–474

Haug GH, Strecker MR (1995) Volcano-tectonic evolution of the Chyulu Hills and implications for the regional stress field in Kenya. Geology 23:165–170

Head JW, Wilson L (1989) Basaltic pyroclastic eruptions: influence of gas release patterns and volume fluxes on fountain structure, and the formation of cinder cones, spatter cones, rootless flows, lava ponds and lava flows. J Volcanol Geoth Res 37:261–271

Heiken GH (1971) Tuff rings: examples from the Fort Rock-Christmas Lake Valley Basin, South-Central Oregon. J Geophy Res 76 (23):5615–5626

Hoblitt RP, Orr TR, Heliker C, Denlinger RP, Hon K, Cervelli PF (2012) Inflation rates, rifts, and bands in a pahoehoe sheet flow. Geosphere 8(1):179–195

James MR, Applegarth LJ, Pinkerton H (2012) Lava channel roofing, overflows, breaches and switching: insights from the 2008-2009 eruption of Mt. Etna Bull Volcanol 74(1):107–117

Jordan SC, Cas RAF, Hayman PC (2013) The origin of a large (>3 km) maar volcano by coalescence of multiple shallow craters: Lake Purrumbete maar, southeastern Australia. J Volcanol Geoth Res 254:5–22

Keating GN, Valentine GA, Krier DJ, Perry FV (2008) Shallow plumbing systems for small-volume basaltic volcanoes. Bull Volc. doi:10.1007/s00445-007-0154-1

Kereszturi G, Németh K (2012) Monogenetic basaltic volcanoes: genetic classification, growth, geomorphology and degradation. In: Németh K (ed), Updates in Volcanology—New Advances in Understanding Volcanic Systems. inTech Open, Rijeka, Croatia, pp 3–88 [http://dx.doi.org/10.5772/51387]

Keszthelyi L, Denlinger R (1996) The initial cooling of pahoehoe flow lobes. Bull Volc 58(1):5–18

Konrad K, Graham DW, Thornber CR, Duncan RA, Kent AJR, Al-Amri AM (2016) Asthenosphere-lithosphere interactions in Western Saudi Arabia: Inferences from $^{3}He/^{4}He$ in xenoliths and lava flows from Harrat Hutaymah. Lithos 248:339–352

Lefebvre NS, White JDL, Kjarsgaard BA (2013) Unbedded diatreme deposits reveal maar-diatreme-forming eruptive processes: standing rocks west, Hopi Buttes, Navajo Nation, USA. Bull Volcanol 75(8)

Lorenz V (1974) On the Formation of Maars. Bulletin Volcanologique 37(2):183–204

Lorenz V (1986) On the growth of maars and diatremes and its relevance to the formation of tuff rings. Bull Volc 48:265–274

Martin U, Németh K (2006) How Strombolian is a "Strombolian" scoria cone? Some irregularities in scoria cone architecture from the Transmexican Volcanic Belt, near Volcan Ceboruco, (Mexico) and Al Haruj (Libya). J Volcanol Geoth Res 155(1–2):104–118

Mattsson HB (2010) Textural variation in juvenile pyroclasts from an emergent, Surtseyan-type, volcanic eruption; the Capelas tuff cone, Sao Miguel (Azores). J Volcanol Geoth Res 189(1–2):81–91

McGuire AV (1988) The mantle beneath the Red Sea margin; xenoliths from western Saudi Arabia. Tectonophysics 150(1–2):101–119

Moore JG, Nakamura K, Alcaraz A (1966) The 1965 eruption of Taal volcano. Science 151(3713):955–960

Moufti MR, Németh K (2013) The intra-continental Harrat Al Madinah Volcanic Field, Western Saudi Arabia: a proposal to establish Harrat Al Madinah as the first volcanic geopark in the Kingdom of Saudi Arabia. Geoheritage 5(3):185–206

Moufti MR, Nemeth K (2014) The White Mountains of Harrat Khaybar, Kingdom of Saudi Arabia. Int J Earth Sci 103(6):1641–1643

Moufti MR, Németh K, El-Masry N, Qaddah A (2013a) Geoheritage values of one of the largest maar craters in the Arabian Peninsula: the Al Wahbah Crater and other volcanoes (Harrat Kishb, Saudi Arabia). Cent Eur J Geosci 5(2):254–271

Moufti RM, Németh K, Murcia H, Lindsay J, El-Masry N (2013b) Geosite of a steep lava spatter cone of the 1256 AD, Al Madinah eruption, Kingdom of Saudi Arabia. Cent Eur J Geosci 5(2):189–195

Moufti M, Németh K, El-Masry N, Qaddah A (2015) Volcanic Geotopes and Their Geosites Preserved in an Arid Climate Related to Landscape and Climate Changes Since the Neogene in Northern Saudi Arabia: Harrat Hutaymah (Hai'il Region). Geoheritage 7 (2):103–118

Murcia H, Németh K, Moufti MR, Lindsay JM, El-Masry N, Cronin SJ, Qaddah A, Smith IEM (2014) Late Holocene lava flow morphotypes of northern Harrat Rahat, Kingdom of Saudi Arabia: Implications for the description of continental lava fields. J Asian Earth Sci 84:131–145

Murcia H, Németh K, El-Masry NN, Lindsay JM, Moufti MRH, Wameyo P, Cronin SJ, Smith IEM, Kereszturi G (2015) The Al-Du'aythah volcanic cones, Al-Madinah City: implications for volcanic hazards in northern Harrat Rahat, Kingdom of Saudi Arabia. Bull Volcanol 77(6)

Németh K (2010) Monogenetic volcanic fields: origin, sedimentary record, and relationship with polygenetic volcanism. In: Canon-Tapia E, Szakacs A (eds) What Is a Volcano?. Geological Society of America, Boulder, Colorado, pp 43–66

Németh K, Cronin SJ (2008) Volcanic craters, pit craters and high-level magma-feeding systems of a mafic island-arc volcano: Ambrym, Vanuatu, South Pacific. In: Thomson K, Petford N (eds), Structure and Emplacement of High-Level Magmatic Systems. Geological Society, London, Special Publications. Geological Society, London, pp 85–99

Németh K, Cronin SJ (2009) Phreatomagmatic volcanic hazards where rift-systems meet the sea, a study from Ambae Island, Vanuatu. J Volcanol Geoth Res 180(2–4):246–258

Németh K, Cronin SJ (2011) Drivers of explosivity and elevated hazard in basaltic fissure eruptions: the 1913 eruption of Ambrym Volcano, Vanuatu (SW-Pacific). J Volcanol Geoth Res 201(1–4):194–209

Németh K, White JDL (2003) Reconstructing eruption processes of a Miocene monogenetic volcanic field from vent remnants: Waipiata Volcanic Field, South Island, New Zealand. J Volcanol Geoth Res 124(1–2):1–21

Németh K, Martin U, Harangi S (2001) Miocene phreatomagmatic volcanism at Tihany (Pannonian Basin, Hungary). J Volcanol Geoth Res 111(1–4):111–135

Németh K, Suwesi KS, Peregi Z, Gulácsi Z, Ujszászi J (2003) Plio/Pleistocene flood basalt related scoria and spatter cones, rootless lava flows, and pit craters, Al Haruj Al Abyad, Libya. Geolines J Geol Inst AS Czech Repub (15):98–103

Németh K, Cronin SJ, Smith IEM, Flores JA (2012a) Amplified hazard of small-volume monogenetic eruptions due to environmental controls, Orakei Basin, Auckland Volcanic Field, New Zealand. Bull Volcanol 74(9):2121–2137

Németh K, Risso C, Nullo F, Smith IEM, Pecskay Z (2012b) Facies architecture of an isolated long-lived, nested polygenetic silicic tuff ring erupted in a braided river system: the Los Loros volcano, Mendoza, Argentina. J Volcanol Geoth Res 239:33–48

Németh K, Moufti MR, El-Masry N, Qaddah A (2013) Magma withdrawal below ground-water table as a trigger to form large maars over high magma discharge rate fissure-fed lava spatter/scoria cones: the Al Wahbah Crater (Harrat Kishb, Saudi Arabia). Abstract Volume of the IAVCEI 2013 [Kagoshima, Japan, 15–19 July 2013], Abstract number:1089-1

Neumann Van Padang M (1963) Catalogue of the active volcanoes of the world including solfatara fields: Part 16, Arabia and the Indian Ocean. International Association of Volcanology

Okubo CH, Martel SJ (1998) Pit crater formation on Kilauea volcano, Hawaii. J Volcanol Geoth Res 86(1–4):1–18

Pallister JS (1985) Reconnaissance geology of the Harrat Hutaymah Quadrangle, sheet 26/42A, Kingdom of Saudi Arabia. Open-File Report-U.S. Geological Survey, pp 82–82

Patrick MR, Orr TR (2012) Rootless shield and perched lava pond collapses at KA << lauea Volcano, Hawaii. Bull Volcanol 74 (1):67–78

Peterson DW, Tilling RI (1980) Transition of basaltic lava from pahoehoe to aa, Kilauea Volcano, Hawaii—Field observations and key factors. J Volcanol Geoth Res 7(3–4):271–293

Riggs N, Carrasco-Nunez G (2004) Evolution of a complex isolated dome system, Cerro Pizarro, central Mexico. Bull Volc 66(4):322–335

Roobol M, Shouman S, Al Solami A (1985) Earth tremors, ground fractures, and damage to buildings at Tabah (27/42C). Saudi Arabian Deputy Ministry for Mineral Resources Technical Record DGMR-TR-05-4

Rossi MJ (1997) Morphology of the 1984 open-channel lava flow at Krafla volcano, northern Iceland. Geomorphology 20(1–2):95–112

Rowland SK, Walker GPL (1990) Pahoehoe and aa in Hawaii - Volumetric flow-rate controls the lava structure. Bull Volc 52 (8):615–628

Rymer H, de Vries BV, Stix J, Williams-Jones G (1998) Pit crater structure and processes governing persistent activity at Masaya Volcano, Nicaragua. Bull Volcanol 59(5):345–355

Sato H (1995) Textural Difference between Pahoehoe and Aa Lavas of Izu-Oshima Volcano, Japan - an Experimental-Study on Population-Density of Plagioclase. J Volcanol Geoth Res 66(1–4):101–113

Schmincke H-U, Fisher RV, Waters A (1973) Antidune and chute and pool structures in the base surge deposits of the Laacher See area, Germany. Sedimentology 20:553–574

Schumacher R, Schmincke H-U (1990) The lateral facies of ignimbrites at Laacher See volcano. Bull Volc 52:271–285

Self S, Keszthelyi L, Thordarson T (1998) The importance of pahoehoe. Annu Rev Earth Planet Sci 26:81–110

Sohn YK, Park KH, Yoon SH (2008) Primary versus secondary and subaerial versus submarine hydrovolcanic deposits in the subsurface of Jeju Island, Korea. Sedimentology 55(4):899–924

Stevenson JA, Mitchell NC, Cassidy M, Pinkerton H (2012) Widespread inflation and drainage of a pahoehoe flow field: the Nesjahraun, Aingvellir, Iceland. Bull Volcanol 74(1):15–31

Stovall WK, Houghton BF, Harris AJL, Swanson DA (2009) Features of lava lake filling and draining and their implications for eruption dynamics. Bull Volc 71(7):767–780

Sumner JM (1998) Formation of clastogenic lava flows during fissure eruption and scoria cone collapse: the 1986 eruption of Izu-Oshima Volcano, eastern Japan. Bull Volc 60:195–212

Tait MA, Cas RAF, Viramonte JG (2009) The origin of an unusual tuff ring of perlitic rhyolite pyroclasts: The last explosive phase of the Ramadas Volcanic Centre, Andean Puna, Salta, NW Argentina. J Volcanol Geoth Res 183(1–2):1–16

Thordarson T, Self S (1993) The Laki (Skaftar-Fires) and Grimsvotn Eruptions in 1783-1785. Bull Volc 55(4):233–263

Thornber CR (1990) Geologic map of Harrat Hutaymah, with petrologic classification and distribution of ultramafic inclusions, Saudi Arabia. U. S. Geological Survey, Reston, VA, Reston, VA, United States (USA), pp 0–1 sheet

Thornber CR (1992) Lithospheric-mantle magmatism associated with rifting of continental terranes; a case study of ultramafic inclusions and mafic alkaline magmas from Harrat Hutaymah, Saudi Arabia. U. S. Geological Survey Circular, pp 18–19

Thornber CR (1993) The petrology, geochemistry and origin of ultramafic inclusions and mafic alkaline volcanics from Harrat Hutaymah, Saudi Arabia. PhD (Doctoral) thesis, Ann Arbor, MI, United States (USA) [unpublished], pp 271–271

Thornber CR (1994) Ultramafic inclusions from Harrat Hutaymah; a record of mantle magmatism beneath north central Arabia. CPRM - Special Publication 1A:434–454

Thornber CR (1988) Hutayman pyroxenite xenoliths; clues to the character of subcontinental mantle during early stage Red Sea rifting. Abs Programs Geol Soc Am 20(7):367

Thornber CR (1991) Hot, cold, wet, and dry Hutaymah ultramafic inclusions; a record of mantle magmatism beneath the Arabian Shield and flanking the Red Sea Rift. Proc Int Kimberlite Conf 5:423–425

Thornber CR, Pallister JS (1985) Mantle xenoliths from northern Saudi Arabia. Eos Trans Am Geophy Union 66(18):393

Ui T, Matsuwo N, Sumita M, Fujinawa A (1999) Generation of block and ash flows during the 1990-1995 eruption of Unzen Volcano, Japan. J Volcanol Geoth Res 89:123–137

Valentine GA, Cortes JA (2013) Time and space variations in magmatic and phreatomagmatic eruptive processes at Easy Chair (Lunar Crater Volcanic Field, Nevada, USA). Bull Volcanol 75(9)

Valentine GA, Gregg TKP (2008) Continental basaltic volcanoes; processes and problems. J Volcanol Geoth Res 177(4):857–873

Valentine GA, White JDL (2012) Revised conceptual model for maar-diatremes; subsurface processes, energetics, and eruptive products. Geology (Boulder) 40(12):1111–1114

Vespermann D, Schmincke H-U, Ballard RD (2000) Scoria cones and tuff rings. Academic Press, San Diego, CA, San Diego, CA, United States (USA)

Vincent P (2008) Saudi Arabia—an environemental overview. Taylor and Francis, London, p 309

White JDL (1991) Maar-diatreme phreatomagmatism at Hopi Buttes, Navajo Nation (Arizona), USA. Bull Volc 53:239–258

White JDL, Houghton BF (2000) Surtseyan and related eruptions. In: Sigurdsson H, Houghton B, McNutt S, Rymer H, Stix J (eds) Encyclopedia of Volcanoes. Academic Press, New York, pp 495–512

White JDL, Ross PS (2011) Maar-diatreme volcanoes: A review. J Volcanol Geoth Res 201(1–4):1–29

Synthesis of the Geoheritage Values of the Volcanic Harrats of Saudi Arabia

5

5.1 Potential Link Between Geoeducational Programs in Various Harrats of Saudi Arabia

Western Saudi Arabia is the home of extensive volcanic fields with hundreds of well-preserved volcanic landforms (Camp and Roobol 1989a, 1992; Camp et al. 1991, 1992; Alwelaie 1994; Bosworth et al. 2005). Due to the arid climate these volcanic landforms are well-preserved and show perfect volcanic edifice architecture that can be utilized in various geoeducational programs (Fig. 5.1). This is an important aspect during the identification and cataloguing of volcanic geoheritage values in the region, which is still considered to be active and future volcanic unrest is expected in areas that are either culturally or economically significant. Volcanoes identified among the harrats of Saudi Arabia contain some of the most common volcanic landforms on Earth and in the Solar System, including scoria cones and spatter cones (Connor and Conway 2000; Walker 2000; Kereszturi and Németh 2012; Németh and Kereszturi 2015). In this respect, it is difficult to argue the "*outstanding universal value*" of the geoheritage of the sites (Badman et al. 2008), using the concept imposed by UNESCO (http://whc.unesco.org/en/criteria/ and https://portals.iucn.org/library/efiles/documents/2008-036.pdf), as they cannot be universally unique if they are among the most common volcanic landforms on Earth. This is a paradox shared by many other geoheritage sites associated with dispersed volcanic fields. This is a major philosophical issue that is the subject of many discussions, including the nomination process for UNESCO World Heritage status of proposed volcanic fields (http://www.chainedespuys-failledelimagne.com/); however, here we demonstrate the unique nature of the Saudi harrats.

The harrats we present here are typical volcanic fields in an intracontinental setting (Németh and Kereszturi 2015), representing the near continuous volcanism in the western margin of the Arabian Peninsula that started nearly 30 million years ago (Yurur and Chorowicz 1998; Ukstins et al. 2002; Shaw et al. 2007; Moufti et al. 2013a; Runge et al. 2014; Wahab et al. 2014) and culminated in several well-defined eruptive stages forming overlapping and spatially well-defined volcanic fields (Camp and Roobol 1989b; Bosworth et al. 2004, 2005). While it is unlikely that any of the volcanic features are individually unique on a global scale, together these fields form a globally unique and significant geological environment, which presents a unique opportunity to see the variety in the eruption styles and nature of intracontinental volcanism. The individual volcanic fields, moreover, can be linked together to provide an extensive region over a continental plate, extending over 2000 km in length in the western margin of the Arabian Peninsula.

Each of the volcanic fields contains a type of volcanic feature that can be identified across many of the fields, including an abundance of volcanic edifices produced by mild explosive eruptions, effusion of transitional type lava flows, dominance of lava-fountain producing eruptions and the formation of aligned lava spatter cones. In this respect, the "*outstanding universal value*" of these volcanoes is not in their individual uniqueness but rather in their abundance and common similarities across a vast area of land in the western margin of the Arabian Peninsula. Hence, the western Arabian volcanic fields together represent a well-defined and dispersed region, where lava-fountain and scoria-cone dominated volcanism is common and produced major landscape-defining elements.

While this common link between the Arabian harrats illustrates the nature of the volcanoes that could be encountered, the detail of their volcanic architecture shows enough uniqueness for them to be treated as individual volcanic geotopes with their unique volcanic geosites. Moreover, there is sufficient geodiversity in the appearance of the spatter and scoria cones across the volcanic fields for visitors to understand the fine details of these volcanic landforms.

A further factor that promotes the harrats being treated as separate geosites is the fact that there are regions in western Arabia that are easier to access than others, and this will determine the extent to which a geoeducation and geotouristic program could be developed in an area. Harrat Rahat is

© Springer International Publishing Switzerland 2016

M.R. Moufti and K. Németh, *Geoheritage of Volcanic Harrats in Saudi Arabia*,
Geoheritage, Geoparks and Geotourism, DOI 10.1007/978-3-319-33015-0_5

Fig. 5.1 Well-preserved volcanic landscape of the Saudi harrats such as tuff rings with intra-crater scoria cone from Harrat Kishb [22° 48′ 20.49″N; 41° 20′ 52.22″E]

particularly well accessible, and it is located beside the culturally and religiously significant city of Al Madinah, making it the ideal location to develop a geoeducational program. Harrat Rahat is also the home of one of the most recent of the known eruption sites of the Arabian Peninsula, which built a chain of scoria and spatter cones and emitted a large volume (about 0.5 km^3) of transitional lava flows (Murcia et al. 2014), making this site a perfect "*knowledge and/or geoeducational hub*" that could then be linked to similar sites in other harrats. The program in this sense would use the 1256 AD eruption sites at Harrat Rahat to "*scale*" the spatter and scoria cones, which visitors then can compare with other similar volcanoes across Arabia (Moufti et al. 2013d). Information leaflets, maps and other electronic resources (e.g., geo-routes) could emphasize the similarities and differences among these volcanic features to show and explain the reason for their commonness and diversity. Spatter cones and scoria cones, while similar to each other and defined by a specific size, volume and eruption style range, differ significantly from each other both within in a single volcanic field and across many volcanic fields (Vespermann and Schmincke 2000; Wolff and Sumner 2000). To characterize this variation in a scientifically correct way is research that has not been conducted yet; however, there are some obvious trends that can be noted. In Harrat Rahat many spatter cones (Fig. 5.2) are of a size that is considered to be large and many scoria cones appear like large (*overgrown*) spatter cones. This indicates that many of the Harrat Rahat's scoria cones erupted

Fig. 5.2 Spatter cone of Harrat Rahat [24° 20′ 42.03″N; 39° 46′ 32.76″E]

through numerous and potentially high lava fountains, producing significant welding, heat radiation and associated molten edifice collapse events (Moufti and Németh 2013; Moufti et al. 2013c). Hence the majority of the preserved scoria cones in the Harrat Rahat are inferred to be dominated by hot temperature magmatic explosive eruptive events. A similar situation can be seen at the Harrat Khaybar, where extensive large scoria cones commonly steep, and align with spatter collar covered cones. It seems that, while no precise study has been conducted yet, the volcanic cones of the harrats are more like large spatter cones than coarse ash- and fine lapilli-dominated cones composed of granular media. An exception from this trend could be the Harrat Hutaymah and Al Birk, where the scoria cones are more non-welded ash and lapilli dominated; however this requires further investigation and in-depth study.

The scoria cones of the harrats occasionally produced eruptions that generated a large volume of fragmented pyroclasts, which dispersed over large areas leaving behind dark ash and lapilli plains. At Harrat Rahat, the 1256 AD eruption left behind a reasonably large ash and lapilli field (Fig. 5.3) that produced a potentially violent Strombolian phase or phases in the eruption of one of the vents (Kawabata et al. 2015), similar to those seen at Harrat Khaybar in association with the Jebel Quidr volcano (Fig. 5.4). These locations represent the most violent eruptive events of mafic explosive volcanism in the region and their potential volcanic hazard implications are huge. These locations, therefore, can be used for future volcanic hazard education programs to demonstrate the consequences of similar eruptions in the future. Similar large violent Strombolian or sub-Plinian style mafic eruption sites are relatively rare across the harrats' volcanoes.

Overall it is clear that together the harrats contain all the known types of scoria and spatter cones and this makes the western Arabian harrats internally and globally unique: they can be defined as carrying high "*outstanding universal value*".

The harrats, however, do not only consist of volcanoes formed by mafic explosive eruptions; there are surprisingly large numbers of silicic eruptive centres. Harrat Rahat hosts one of the most diverse varieties of volcanoes in the region, ranging from maars formed due to magma and water explosive interactions to fully developed small-to-medium

Fig. 5.3 Ash plain 1256 AD site [24° 20′ 51.18″N; 39° 46′ 34.32″E]

Fig. 5.4 Ash plain at Harrat Khaybar near Jebel Quidr [25° 43′ 44.41″N; 39° 58′ 9.39″E]

calderas that erupted due to the expansion of magmatic volatiles of silicic (trachytic) magmas (Camp and Roobol 1989a; Moufti and Németh 2013). Harrat Rahat in this respect carries large "*outstanding universal value*" as it can demonstrate the diversity of explosive eruption sites from purely externally to dominantly internally driven explosive magma fragmentations, producing diverse types of volcanic craters. Similar diversity of explosive craters and silicic volcanism is known from the Harat Khaybar (Moufti and Németh 2014) and Harrat Kishb (Grainger 1996; Moufti et al. 2013b; Wahab et al. 2014). Harrat Khaybar represents a compositionally similar scenario to Harrat Rahat, while Harrat Khisb represents a more alkali-dominated volcanism with chemically different eruptive products (Camp and Roobol 1989a; Camp et al. 1991, 1992). These three harrats, however, are unique in the way they demonstrate a very broad chemical variety of eruptive products, which form distinctly different volcanic edifices ranging from pure explosion craters to complex caldera-like features and/or silicic lava domes. Harrat Rahat and Harrat Khaybar, in addition, show numerous instances of the co-location of mafic and silicic eruption sites, forming complex and compound volcanic edifices (Fig. 5.5). These have individual characteristics and hence are easy to define as volcanic geotopes with unique volcanic geosites. The Harrats Khaybar, Rahat and Khisb together exhibit a complex and mature stage in the evolution of dispersed volcanism, and also links to polygenetic volcanism; therefore, together they carry "*outstanding universal value*" beside their huge aesthetic value.

Harrat Hutaymah is an example of the most typical of small-volume mafic volcanic fields with numerous maar volcanoes (Pallister 1985; Thornber 1990; Moufti et al. 2015). There are maar volcanoes that are filled with post-eruptive debris and there are those that function as large open craters today (Fig. 5.6). This diversity of the maar craters preserved across Harrat Hutaymah makes this field of "*outstanding universal value*". While Harrat Hutaymah is clearly the volcanic field in the Arabian Peninsula with the greatest abundance of maar volcanoes, there are other sites in Arabia that have large and very well-preserved young maar volcanoes. Harrat Khisb is one such example, with its single large maar crater (Al Wahbah) and several smaller maars partially buried (Fig. 5.7). In contrast, Harrat Al Birk is a volcanic field where maar craters are commonly hidden by aeolian sand cover and potentially many of the sites have

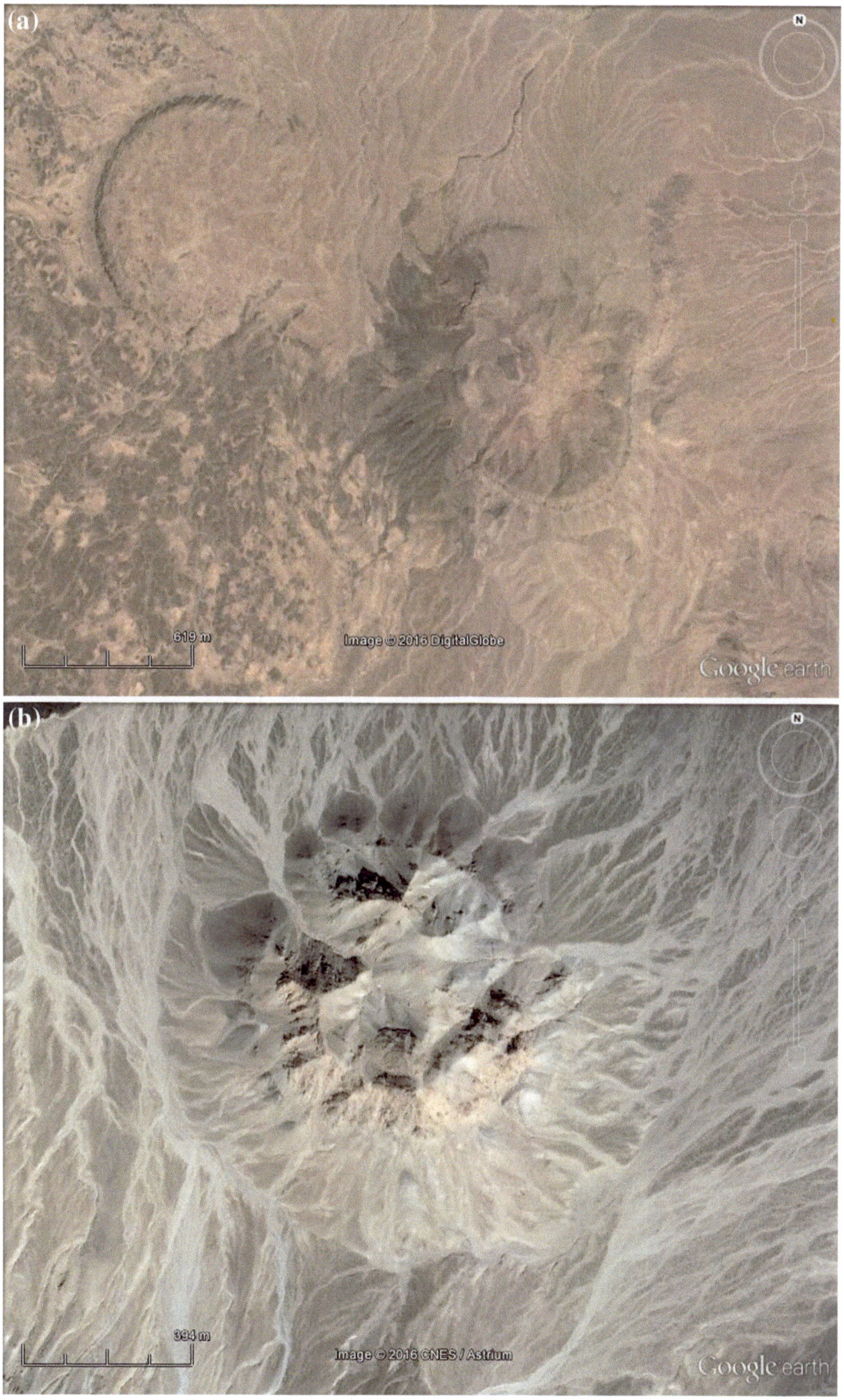

Fig. 5.5 Complex silicic volcanoes of Harrat Rahat (**a** 24° 5′ 38.59″N; 39° 55′ 32.47″E) and Harrat Khaybar (**b** 25° 40′ 29.67″N; 39° 58′ 47.31″E) (Google Earth Image)

been misinterpreted and/or even missed in previous work (Fig. 5.8). The variety of maar craters and their infill status across the harrats of western Arabia is huge and this can potentially serve a well-linked educational program to demonstrate how a terrestrial crater can be filled. Harrat Hutaymah is also a location where maar crater rim deposits are very well exposed and thus is an ideal site for demonstrating the potential deposits a maar-forming eruption could

Fig. 5.6 Harrat Hutaymah crater [26° 59′ 39.62″N; 42° 15′ 6.37″E]

Fig. 5.7 Al Wahbah maar with wide and deep crater [22° 53′ 37.65″N; 41° 7′ 59.97″E]

contribute to the surroundings. While individual maar volcanoes are not unique in the harrats of Arabia, their abundance in some harrats and rarity in others is significant, as is their ability to demonstrate the link between the external and internal controlling parameters responsible for the eruption of small-volume basaltic volcanoes (Németh et al. 2012; Németh and Kereszturi 2015). The harrats of Arabia, in this respect, carry also "*outstanding universal value*".

The most common volcanic landscape-forming elements of each harrat are the vast lava flows (Fig. 5.9). They occur in each of the harrats and share the same typical transitional (from pahoehoe to aa) lava flow surface textures (Murcia et al. 2014). The abundance of such lava surface textures perhaps diminishes the unique image of the lava flows and seemingly degrades their geoheritage value, however, again their significance and global uniqueness lay in their abundance in the region. A geoeducational program highlighting this fact could be significant in volcanic hazard education in the region, to help appreciate the type of lava flows future eruptions would produce.

Overall the harrats of Arabia show similar architectural features, similar volcanic edifices and similar lava flow

Fig. 5.8 Hidden maar in Al Birk near the town of Sabya on a Google Earth image. The maar crater has been recognized beneath an intra-crater lava flow (centre of image) that partially covered by aeolian sediments [17° 10′ 58.59″N; 42° 44′ 4.03″E]

textures. These similarities are the key to link these volcanic fields and demonstrate that together the region contains nearly every possible feature one can imagine relevant to dispersed intra-continental volcanism. The real "*outstanding universal value*" of the harrats of Arabia lies in the abundance of its volcanic features, which formed by every known internally and externally controlled geological forces.

5.2 Potential Links of the Identified Volcanic Geoheritage Sites in Saudi Arabia with Others Around the World

The harrats of Arabia can easily be linked to other dispersed volcanic fields and their corresponding geoeducational programs. The harrats of Arabia are unique as one of the most diverse regions worldwide, in terms of the preserved volcanic landforms. This is the key to be able to link the Saudi harrats with other intra-continental volcanic fields on Earth.

The length of volcanic activity in each of the Arabian harrats is in the range of several millions of years, which makes these volcanic fields fairly mature volcanic regions. The mature nature of the volcanic fields is expressed through the common presence of complex volcanoes with compound architecture and a general overlapping and amalgamated volcanic stratigraphy. Harrat Rahat, for instance, was active over the past 10 million years and produced distinct volcanic stratigraphy horizons with an aggregated thickness of volcanic products over 100 m in inter-cone regions (Moufti et al. 2013a). Such long-lived and productive volcanic fields are also known elsewhere and this provides a golden opportunity to develop a global network of similar volcanic regions. Harrat Rahat has numerous analogues in East

Fig. 5.9 Vast lava flow fields of Harrat Khaybar [25° 42′ 41.37″N; 39° 58′ 27.07″E]

Fig. 5.10 Overview of the Chyulu volcanic field (Kenya, East Africa) from the south [2° 55′ 4.68″S; 38° 0′ 43.33″E]

Africa, such as in regions in the Chyulu volcanic field (Novak et al. 1997; Spath et al. 2001; Sakkas et al. 2002) that are dominated by stacks of thick lava flows and volcanic cones range from very small volume edifices to complex and compound ones (Fig. 5.10). Many central Anatolia volcanic fields share similar architecture to Harrat Rahat, and the chemical diversity of the Anatolian volcanic fields can be as broad as has been documented from Harrat Rahat (Aydin et al. 2014; Seghedi et al. 2015). Probably the most similar volcanic fields to Harrat Rahat and Harrat Khaybar, where chemical diversity of the eruptive products and the variety of volcanic eruptions styles and hence volcanic landforms are huge, are those of the Transmexican Volcanic belt (Wallace and Carmichael 1999; Petrone 2010; Agustin-Flores et al. 2011; Guilbaud et al. 2012), the Colorado Plateau volcanic regions (Elston and Wohletz 1987; Arculus and Gust 1995; Valentine and Cortes 2013), and Chaine-de-Puys in France (Morel et al. 1992; Woodland and Jugo 2007). These regions share the common fact with Harrat Rahat, Harrat Khaybar and even with Harrat Kishb that they are also the products of volcanism fed by both mafic and silicic magmas. Somewhat similar to these Saudi volcanic fields are the extensive volcanic regions of Victoria and South Australia, at least from the longevity and/or the total eruptive volume of the volcanism produced there and in Arabia (Cas 1989; Hare and Cas 2005; Boyce et al. 2014).

Interestingly, Harrat Hutaymah is the only volcanic field in Arabia to show close similarity to the volcanic fields across Central Europe from the Eifel (Schmincke et al. 1983; Bitschene and Schueller 2011) through the Eger Graben (Cajz et al. 2009) to the Pannonian Basin (Martin and Németh 2004), and the Patagonian basaltic fields mostly in Argentina (Pankhurst et al. 1998; D'Orazio et al. 2001; Massaferro et al. 2006). Harrat Hutaymah, therefore, could easily be linked in geoeducational perspective to those fields. Other harrats, such as the Harrat Al Birk, show more similarity through the dominance of lava flows, and the concentrated vent distribution to those volcanic fields located in North Africa, such as Al Haruj in Libya (Németh et al. 2002; Cvetkovic et al. 2010; Bardintzeff et al. 2012).

Overall there are numerous connection points between the harrats of Arabia and other volcanic fields globally. Many of the global analogues are at an advanced stage in the research and characterisation of their volcanic geoheritage values, making it a logical step to link the Saudi harrats to them. The large number and the great diversity of these dispersed volcanic fields and their volcanoes are the keys to view these monogenetic volcanic fields as important contributors to the total volcanism on Earth. The Saudi harrats together cover every known aspect of monogenetic volcanism, making them a key location to be connected to global geo-education projects on dispersed volcanism.

5.3 Potential Global Networking and the Role of the Saudi Volcanic Harrats

The current project to propose the Chaine-de-Puys volcanic region along the Limagne Fault line in France as a UNESCO World Heritage Site highlighted the pitfalls of the evaluation process and criteria UNESCO imposes on potential candidates. The nature of monogenetic volcanic fields means that they contain very common volcanic landforms and individually it would be very difficult to argue that they carry a globally significant "*outstanding universal value*". While a simple scoria cone is not exciting from a volcanic landscape perspective or from its potential to provide vital new information for the understanding of volcanic eruptions, if they exist in large groups and show great variety in their volcanic landforms, they together can form a "*critical mass*" of cones that can be justified as carrying "*outstanding universal value*". The Saudi harrats in this respect can be viewed as globally significant sites along many different geological lines of argument, such as the abundance of diverse types of scoria cones and spatter cones, the compositional variety and great volume of their lava flows. The Saudi harrats are also located in an arid climate, where erosion is slow and the preservation potential of volcanic features is great, alongside good exposures. While the harsh environmental conditions where the Saudi harrats are located does not make them an easy volcano tourist destination, together they form a vital part of a global network to provide analogue and well-exposed text-book-style examples to demonstrate volcanic processes associated with monogenetic volcanoes, and the rock textures and edifice architecture of small-volume volcanoes. The Saudi harrats, therefore, could be linked to a global network of dispersed volcanic fields to provide good examples for features that may not be well exposed elsewhere, demonstrate specific volcano types that may be difficult to fit into a big picture at another volcanic field or to provide an alternative example to demonstrate key volcanic features in order to keep the diversity of information high in any geo-educational program designed elsewhere.

Overall it is expected that a dispersed volcanic field somewhere on Earth will, one day, be granted World Heritage Status, as they represent the most common and most typical type of volcanism in the entire Solar System. The most advanced project to achieve this status has been prepared by the Chaine-de-Puys volcanic region in France. The Chaine-de-Puys volcanic region is also expected and designed to act as a hub for volcanic field geoheritage projects on a global scale. The Saudi harrats are also perfectly linkable to any volcanic fields on Earth with similar geoheritage values.

It would be ideal for global geo-education program if modern, recent and ancient volcanic fields and their volcanic geoheritage values would be linked together. In such a situation, a maar-diatreme volcanic field, such as the eroded ancient region of the Hopi Butte where the diatremes are exhumed, could be linked to Harrat Hutaymah, where the maar volcanic landforms are well-preserved and diatremes are deep beneath their maars' crater floors. The arid nature of the Arabian harrats also can be used to demonstrate the scale and effect of climate change over millions of years, especially by showing the presence of maar volcanoes that were clearly erupted due to explosive magma and water interaction in the past, such as those at Harrat Hutaymah. Other young, present day dry maars, such as the Al Wahbah maar in Harrat Kishb, could be linked to maar geoheritage sites, such as the Messel UNESCO World Heritage site (http://whc.unesco.org/en/list/720) to show what a maar crater floor would look like if water were to be taken away. In a similar way, most of the maar craters identified among the Saudi harrats can be associated with other maar fields with similar volcanic architecture, such as the Valley del Santiago dry maars in central Mexico (Molina Garza et al. 2000; Cano-Cruz and Carrasco-Nunez 2008). These "*bilateral*" links are thought to be essential to develop a truly global network of geoheritage sites of volcanic fields.

The harrats of Arabia, however, can provide a very unique connection point to other dispersed volcanic fields if we consider their long-lived eruption history and common mature geological features. These include having the full spectrum of geochemical lineages from mafic to more silicic version of magmas (e.g., basanite to phonolite or basalt to trachyte) and various small-to-medium volume volcanic landforms, such as simple and single explosion craters to lava dome complexes and associated block-and-ash flow fans linked to local small caldera-like features. These features mark the maturity of a volcanic field's evolution and a global effort to link such sites together, especially those that have had intensive geoheritage study and catalogue programs, is a very desirable process. The Harrat Rahat, Harrat Khaybar and Harrat Khisb are especially good candidates for such global links. Harrat Rahat, with its long-lived and overlapping volcanic fields and centrally located concentration of silicic volcanic centres, show similarity to volcanic fields in the western US (e.g. San Francisco Volcanic Field) (Conway et al. 1998) and many in Sudan (e.g. Meidob Volcanic Field) (Franz et al. 1997). The dorsal ridge like distribution of vents and lava flow fields at Harrat Rahat shows similarities to the Chaine-de-Puys volcanic field in central France (Anonymous 2012; de Vries et al. 2014) and Chyulu Hills in Kenya (Novak et al. 1997). Harrat Khaybar can also be linked to the above mentioned volcanic fields; however the dorsal nature of the vent distribution in Harrat Khaybar is not the same at Harrat Rahat. Harrat Khaybar has more evenly distributed vents, similar to those in Pinacate in Sonora, Mexico (Gutmann and Sheridan 1978; Gutmann 2002). However Harrat Khaybar is more silicic and the abundance of silicic eruptive centres and their eruptive products is greater than in Pinacate. In this respect Harrat Khaybar shows more similarity to volcanic fields in Patagonia, where mafic volcanic fields are commonly associated with medium sized central vents of silicic volcanoes (Massaferro et al. 2006). Harrat Khisb differs from Harrat Rahat and Khaybar, being more alkali in its eruptive products, and it shows great similarity in its volcanic geoheritage values to those fields located in the Central European Volcanic Province. The link between Harrat Khisb and the Eifel in Germany is particularly important, as the Vulkaneifel itself is a Global Geopark (http://www.geopark-vulkaneifel.de/index.php/en/) with a strong emphasis on its volcanic geoheritage (Bitschene and Schueller 2011). Other Central European volcanic geoparks, such as the Bakony-Balaton Highland Geopark (http://www.geopark.hu/en/) or the Nograd/Novohrad Geopark (http://www.nogradgeopark.eu/), are therefore also natural partners to be associated with Harrat Khisb and co-develop geoeducational programs that reference the other sites.

5.4 Outlook

Overall, the Arabian volcanic geoheritage sites together form a globally significant location of intra-plate volcanism, and they carry huge geoheritage values, which are ready to be linked to a global network of volcanic geoheritage programs, including geoeducation, geotourism and geoconservation. While the logistical issues to visit the Arabian sites, including natural and geopolitical obstacles, need to be carefully managed and resolved, in the long term, the Arabian volcanic geoheritage sites have a huge role in the future to act as a natural laboratory that scientists and the general public can enjoy and appreciate equally. Future geoheritage research and work need to explore these opportunities.

Volcanic geoheritage studies in the Arabian Peninsula have been concentrated over the last 5 years. This research is largely an offspring from other concentrated volcanic hazard related work in the region. Geoheritage studies in the region are therefore in their infancy and further work is required before a robust and functioning geoheritage program can be established that is equally accepted by end users, government officials, nature conservationists, geotour operators, and researchers.

While at this stage the identification of geosites and the evaluation of their geoheritage values from the local to the global scale are most advanced in the Harrat Rahat, further research is needed to reach the same detail of understanding the geoheritage values of other harrats. The Harrat Khaybar and Harrat Hutaymah geosite cataloguing process is reasonably well-advanced, but further research is needed to make the study complete. Such studies are also needed to open the door toward other geoheritage aspects of these fields, including non-volcanic heritage aspects, as well as links to other cultural heritage information. Similar research is still in a very preliminary stage in other harrats, and only fragmented research has been conducted at Harrat Khisb and Harrat Al Birk. These two harrats have also enormous geoheritage potential, and by their locations can provide a more complete volcanic geoheritage picture for the Arabian Peninsula. Beside the documented and described volcanic geoheritage sites of the harrats of Rahat, Khaybar, Hutaymah, Khisb and Al Birk, there are other harrats in the region that have not been studied at all, and they potentially hold some other key sites and/or other key aspects that could be added to the portfolio of the harrats of Arabia. Among these unstudied sites is the Harrat Lunayyir, which is located near to the Red Sea coast, and has been identified as a target for scientific research since the seismic unrest period in 2009, which was suspected to be the result of a failed eruption caused by intruding magma to shallow level (Pallister et al. 2010; Hansen et al. 2013; Mukhopadhyay et al. 2013; Zobin et al. 2013; Koulakov et al. 2014). This field also has well preserved volcanic landforms, which are suspected to be young and have high geoheritage values (Coleman and Gregory 1983). There are also large volcanic fields located in the northern margin of the Arabian Peninsula, such as the Harrat ash Shamah, which is considered to be the largest volcanic field in the region by territory (Coleman and Gregory 1983; Camp and Roobol 1992). While this region is located in the territory between Syria and Saudi Arabia and considered to be difficult to access, for future research it is an area where a preliminary geoheritage evaluation should be planned (Snyder et al. 1993; Almalabeh 1994; Nasir 1995; Nasir and Safarjalani 2000; Krienitz et al. 2007; Al-Safarjalani et al. 2009; Krienitz and Haase 2011). There are older volcanic fields in Arabia, such as the Harrat Uwayrid and Harrat ar Raha (Coleman and Gregory 1983; Lange and Anonymous 1986; Blusztajn et al. 1995; Kaliwoda et al. 2005, 2007, 2008), both located nearby to a UNESCO World Heritage site (Al-Hijr Archaeological Site (Madâin Sâlih)—http://whc.unesco.org/en/list/1293). This site was selected on the basis of its cultural value; therefore, the volcanic fields could potentially be linked to those culturally important sites. In addition, as these harrats are older, the volcanic landforms are eroded, and deeper zones of their volcanoes are exposed, for example their volcanic conduits (Kereszturi and Németh 2012), providing a potential connection point to modern and well-preserved volcano geosites. Such geoheritage research would make the Arabian harrats together an ideal set of intra-continental volcanic fields, where visitors could see and learn the volcano architecture from the internal structures of volcanoes to their volcanic morphology. This could classify the Arabian harrats as among the most complex and intact volcanic regions where one can learn to understand intraplate monogenetic volcanoes.

Perhaps the ultimate goal is to record the volcanic geoheritage values of each harrats very precisely following the internationally recognized scientific method (Vujičić et al. 2011; Zunino et al. 2012; 박준형 and Cheong 2012; Moufti et al. 2013b; Prieto 2013; Farsani et al. 2014; Lim 2014; Melelli 2014; Fernandez et al. 2014; Rocha et al. 2014; Tomic and Bozic 2014; Brilha 2015; Francesco et al. 2015; Strba et al. 2015). Such work can then provide the scientific basis for local, regional and international councils to organise geoparks of various levels and scales that can together offer one of the largest and most intact intraplate volcanic field-oriented geoeducational programs on Earth with strong geotouristic and geoconservation links.

References

Agustin-Flores J, Siebe C, Guilbaud MN (2011) Geology and geochemistry of Pelagatos, Cerro del Agua, and Dos Cerros monogenetic volcanoes in the Sierra Chichinautzin volcanic field, south of Mexico City. J Volcanol Geoth Res 201(1–4):143–162

Almalabeh A (1994) Geochemistry of 2 volcanic cones from the intracontinental plateau basalt of Harra El-Jabban, Ne-Jordan. Geochem J 28(6):517–540

Al-Safarjalani A, Nasir S, Fockenberg T, Massonne H-J (2009) Chemical composition of crustal xenoliths from southwestern Syria: Characterization of the upper part of the lower crust beneath the Arabian plate. Chem Erde 69(4):359–375

Alwelaie AN (1994) Protected areas in Saudi Arabia: Sustainable use of natural resources. GeoJournal 34(4):383–392

Anonymous (2012) Synthesis: Noimination for inscription on the world heritage list—Tectono-volcanic ensemble of the Chaine des Puys and the Limagne fault, France. Conseil général du Puy-de-Dome, Clermont-Ferrand, France, p 44

Arculus RJ, Gust DA (1995) Regional petrology of the San-Francisco volcanic field, Arizona, USA. J Petrol 36(3):827–861

Aydin F, Schmitt AK, Siebel W, Sonmez M, Ersoy Y, Lermi A, Dirik K, Duncan R (2014) Quaternary bimodal volcanism in the Nigde volcanic complex (Cappadocia, central Anatolia, Turkey): Age, petrogenesis and geodynamic implications. Contrib Miner Petrol 168(5):1–24

Badman T, Bomhard B, Fincke A, Langley J, Rosabal P, Sheppard D (2008) Outstanding universal value: Standards for natural world heritage. In: International union for conservation of nature and natural resources protected areas programme—world heritage studies, Gland, Switzerland, pp 52

Bardintzeff JM, Deniel C, Guillou H, Platevoet B, Telouk P, Oun KM (2012) Miocene to recent alkaline volcanism between Al Haruj and Waw an Namous (southern Libya). Int J Earth Sci 101(4): 1047–1063

Bitschene P, Schueller A (2011) Geo-education and geopark implementation in the Vulkaneifel European Geopark. GSA Field Guide 22:29–34

Blusztajn J, Hart SR, Shimizu N, McGuire AV (1995) Trace-element and isotopic characteristics of spinel peridotite xenoliths from Saudi-Arabia. Chem Geol 123(1–4):53–65

Bosworth W, Burke K, Anonymous (2004) Tectonic evolution of the Red Sea; integration and synthesis of outcrop, subsurface and radiometric data. Int Geol Congr Abs Congr Geol Int Res 32(Part 1):627

Bosworth W, Huchon P, McClay K (2005) The Red Sea and Gulf of Aden basins. J Afr Earth Sci 43(1–3):334–378

Boyce JA, Keays RR, Nicholls IA, Hayman P (2014) Eruption centres of the Hamilton area of the Newer Volcanics Province, Victoria, Australia: Pinpointing volcanoes from a multifaceted approach to landform mapping. Aust J Earth Sci 61(5):735–754

Brilha J (2015) Inventory and quantitative assesment of geosites and geodiversity sites: A review. Geoheritage. doi:10.1007/s12371-014-0139-3

Cajz V, Rapprich V, Erban V, Pecskay Z, Radon M (2009) Late Miocene volcanic activity in the Ceske stredohori Mountains (Ohre/Eger Graben, northern Bohemia). Geol Carpath 60(6): 519–533

Camp VE, Roobol MJ (1989a) The Arabian continental Alkali Basalt province. 1. Evolution of Harrat-Rahat, Rahat, Kingdom-of-Saudi-Arabia. Geol Soc Am Bull 101(1):71–95

Camp VE, Roobol MJ (1989b) The Arabian continental alkali basalt province; Part I, evolution of Harrat Rahat, Kingdom of Saudi Arabia; with Suppl. Data 89-04. Geol Soc Am Bull 101(1):71–95

Camp VE, Roobol MJ (1992) Upwelling asthenosphere beneath Western Arabia and its regional implications. J Geophys Res Solid Earth 97(B11):15255–15271

Camp VE, Roobol MJ, Hooper PR (1991) The Arabian Continental Alkali Basalt Province. 2. Evolution of Harrats Khaybar, Ithnayn, and Kura, Kingdom of Saudi-Arabia. Geol Soc Am Bull 103(3): 363–391

Camp VE, Roobol MJ, Hooper PR (1992) The Arabian Continental Alkali Basalt Province. 3. Evolution of Harrat Kishb, Kingdom of Saudi-Arabia. Geol Soc Am Bull 104(4):379–396

Cano-Cruz M, Carrasco-Nunez G (2008) Evolution of a rhyolitic explosion crater (maar): Hoya de Estrada, Valle de Santiago volcanic field Guanajuato, Mexico. Rev Mex Cienc Geol 25(3): 549–564

Cas RAF (1989) Intraplate volcanism in eastern Australia and New Zealand. Cambridge Univ. Press, Cambridge, Cambridge, United Kingdom (GBR)

Coleman RC, Gregory R (1983) Cenozoic volcanic rocks. U.S. Geological Survey Professional Paper: 287

Connor CB, Conway FM (2000) Basaltic volcanic fields. In: Sigurdsson H (ed) Encyclopedia of Volcanoes. Academic Press, San Diego, pp 331–343

Conway FM, Connor CB, Hill BE, Condit CD, Mullaney K, Hall CM (1998) Recurrence rates of basaltic volcanism in SP Cluster, San Francisco volcanic field, Arizona. Geology 26:655–658

Cvetkovic V, Toljic M, Ammar NA, Rundic L, Trish KB (2010) Petrogenesis of the eastern part of the Al Haruj basalts (Libya). J Afr Earth Sci 58(1):37–50

de Vries BW, Marquez A, Herrera R, Granja Bruna JL, Llanes P, Delcamp A (2014) Craters of elevation revisited: Forced-folds, bulging and uplift of volcanoes. Bull Volc 76(11):1–20

D'Orazio M, Agostini S, Innocenti F, Haller MJ, Manetti P, Mazzarini F (2001) Slab window-related magmatism from southernmost South America: the Late Miocene mafic volcanics from the Estancia Glencross area (similar to 52 degrees S, Argentina-Chile). Lithos 57 (2–3):67–89

Elston WE, Wohletz KH (1987) Quaternary basalt fields of west-central New Mexico; McCartys pahoehoe flow, Zuni Canyon aa flow, Auni ice cave, Bandera Crater, and Zuni Salt Lake maar. Geol. Soc. Am, Boulder, CO, United States (USA)

Farsani NT, Coelho COA, Costa CMM, Amrikazemi A (2014) Geo-knowledge management and geoconservation via geoparks and geotourism. Geoheritage 6(3):185–192

Fernandez MP, Lagar Timon D, Garcia Marin R (2014) Geosites inventory in the geopark Villuercas-Ibores-Jara (Extremadura, Spain): A proposal for a new classification. Geoheritage 6(1):17–27

Francesco G, Mario B, Giuseppe P (2015) Environmental geology applied to geoconservation in the interaction between geosites and linear infrastructures in South-Eastern Italy. Geoheritage 7(1):33–46

Franz G, Breitkreuz C, Coyle DA, El Hur B, Heinrich W, Paulick H, Pudlo D, Smith R, Steiner G (1997) The alkaline Meidob volcanic field (Late Cenozoic, northwest Sudan). J Afr Earth Sci 25(2):263–291

Grainger DJ (1996) Al Wahbah volcanic explosion crater, Saudi Arabia. Geology Today January–February: 27–30

Guilbaud M-N, Siebe C, Layer P, Salinas S (2012) Reconstruction of the volcanic history of the Tacambaro-Puruaran area (Michoacan, Mexico) reveals high frequency of Holocene monogenetic eruptions. Bull Volc 74(5):1187–1211

Gutmann JT (2002) Strombolian and effusive activity as precursors to phreatomagmatism: Eruptive sequence at maars of the Pinacate volcanic field, Sonora, Mexico. J Volcanol Geoth Res 113(1–2):345–356

Gutmann JT, Sheridan M (1978) Geology of the pinacate volcanic field. Arizona Bur. Geol Min Tech. Spec. Pap 2:47–59

Hansen SE, DeShon HR, Moore-Driskell MM, Al-Amri AMS (2013) Investigating the P wave velocity structure beneath Harrat Lunayyir, northwestern Saudi Arabia, using double-difference tomography and earthquakes from the 2009 seismic swarm. J Geophys Res Solid Earth 118(9):4814–4826

Hare AG, Cas RAF (2005) Volcanology and evolution of the Werribee Plains intraplate, basaltic lava flow-field, Newer Volcanics Province, southeast Australia. Aust J Earth Sci 52(1):59–78

Kaliwoda M, Altherr R, Anonymous (2005) Mantle xenoliths from Harrat Uwayrid (Saudi Arabia); metasomatic phenomena, thermal history and Li-Be-B characteristics. Ofioliti 30(2):191–192

Kaliwoda M, Altherr R, Meyer H-P (2007) Composition and thermal evolution of the lithospheric mantle beneath the Harrat Uwayrid, eastern flank of the Red Sea Rift (Saudi Arabia). Lithos 99(1–2):105–120

Kaliwoda M, Ludwig T, Altherr R (2008) A new SIMS study of Li, Be, B and delta(7) Li in mantle xenoliths from Harrat Uwayrid (Saudi Arabia). Lithos 106(3–4):261–279

Kawabata E, Cronin SJ, Bebbington MS, Moufti MRH, El-Masry N, Wang T (2015) Identifying multiple eruption phases from a compound tephra blanket: An example of the AD1256 Al-Madinah eruption, Saudi Arabia. Bull Volc 77(1):1–13

Kereszturi G, Németh K (2012) Monogenetic basaltic volcanoes: Genetic classification, growth, geomorphology and degradation. In: Németh K (ed) Updates in volcanology—New advances in understanding volcanic systems in Tech Open, Rijeka, Croatia, pp 3–88 http://dx.doi.org/10.5772/51387

Koulakov I, El Khrepy S, Al-Arifi N, Sychev I, Kuznetsov P (2014) Evidence of magma activation beneath the Harrat Lunayyir basaltic field (Saudi Arabia) from attenuation tomography. Solid Earth 5 (2):873–882

Krienitz M-S, Haase KM (2011) The evolution of the Arabian lower crust and lithospheric mantle; geochemical constraints from southern Syrian mafic and ultramafic xenoliths. Chem Geol 280(3–4):271–283

Krienitz MS, Haase KM, Mezger K, Shaikh-Mashail MA (2007) Magma genesis and mantle dynamics at the Harrat Ash Shamah volcanic field (Southern Syria). J Petrol 48(8):1513–1542

Lange U, Anonymous (1986) Basaltverwitterung und Satellitenbildinterpretation des Harrat Uwayrid (Saudi-Arabien); Teilprojekt C6 (A6). Sonderforschungsbereich 108: Spannung und Spannungsumwandlung in der Lithosphaere, Universitaet Karlsruhe, Berichtsband 1984(B):1161–1176

Lim K (2014) A study of geotourism growth through recognition of geoeducation and geoconservation for the geoheritage. J Tour Leis Res 26(6):43–59

Martin U, Németh K (2004) Mio/Pliocene phreatomagmatic volcanism in the western Pannonian Basin. Geol Hung Ser Geol 26:192

Massaferro GI, Haller MJ, D'Orazio M, Alric VI (2006) Sub-recent volcanism in Northern Patagonia: A tectonomagmatic approach. J Volcanol Geoth Res 155(3–4):227–243

Melelli L (2014) Geodiversity: A new quantitative index for natural protected areas enhancement. GeoJ Tour Geo 13(1):27–37

Molina Garza RS, Boehnel H, Perez R, Escanero A, Antretter M (2000) The potential of maar lakes from Valle de Santiago volcanic field, Mexico, for geomagnetic secular variation and paleoclimate studies; preliminary results. Terra Nostra (Bonn) 2000–2006:144–148

Morel JM, Gourgaud A, Vincent PM (1992) A climactic hydromagmatic Eruption in the Aiguiller Massif (Mont-Dore Strato-Volcano, French Massif-Central). Bull Soc Geol Fr 163(5):625–634

Moufti MR, Németh K (2013) The intra-continental Al Madinah volcanic field, Western Saudi Arabia: A proposal to establish Harrat Al Madinah as the first volcanic Geopark in the Kingdom of Saudi Arabia. Geoheritage 5(3):185–206

Moufti MR, Németh K (2014) The White Mountains of Harrat Khaybar, Kingdom of Saudi Arabia. Int J Earth Sci 103(6):1641–1643

Moufti MR, Moghazi AM, Ali KA (2013a) Ar-40/Ar-39 geochronology of the Neogene-Quaternary Harrat Al-Madinah intercontinental volcanic field, Saudi Arabia: Implications for duration and migration of volcanic activity. J Asian Earth Sci 62:253–268

Moufti MR, Németh K, El-Masry N, Qaddah A (2013b) Geoheritage values of one of the largest maar craters in the Arabian Peninsula: The Al Wahbah Crater and other volcanoes (Harrat Kishb, Saudi Arabia). Centr Eur J Geosci 5(2):254–271

Moufti MR, Németh K, Murcia H, Lindsay JM (2013c) The 1256 AD Al Madinah historic eruption geosite as the youngest volcanic chain in the Kingdom of Saudi Arabia. Int J Earth Sci 102(4):1069–1070

Moufti RM, Németh K, Murcia H, Lindsay J, El-Masry N (2013d) Geosite of a steep lava spatter cone of the 1256 AD, Al Madinah eruption, Kingdom of Saudi Arabia. Cent Eur J Geosci 5(2):189–195

Moufti M, Németh K, El-Masry N, Qaddah A (2015) Volcanic geotopes and their geosites preserved in an arid climate related to landscape and climate changes since the neogene in Northern Saudi Arabia: Harrat Hutaymah (Hai'il Region). Geoheritage 7(2):103–118

Mukhopadhyay B, Mogren S, Mukhopadhyay M, Dasgupta S (2013) Incipient status of dyke intrusion in top crust—Evidences from the Al-Ays 2009 earthquake swarm, Harrat Lunayyir, SW Saudi Arabia. Geo Nat Hazards Risk 4(1):30–48

Murcia H, Németh K, Moufti MR, Lindsay JM, El-Masry N, Cronin SJ, Qaddah A, Smith IEM (2014) Late Holocene lava flow morphotypes of northern Harrat Rahat, Kingdom of Saudi Arabia: Implications for the description of continental lava fields. J Asian Earth Sci 84:131–145

Nasir S (1995) Cr-Poor Megacrysts from the Shamah Volcanic field, Northwestern part of the Arabian plate. J Afr Earth Sci 21(3):349–357

Nasir S, Safarjalani A (2000) Lithospheric petrology beneath the northern part of the Arabian Plate in Syria: Evidence from xenoliths in alkali basalts. J Afr Earth Sci 30(1):149–168

Németh K, Kereszturi G (2015) Monogenetic volcanism: Personal views and discussion. Int J Earth Sci 104(8):2131–2146

Németh K, Suwesi SK, Peregi Z, Gulacsi Z, Ujszaszi J, Anonymous (2002) Plio/Pleistocene flood basalt related scoria and spatter cones, rootless lava flows, and pit craters, Al Haruj al Abiyad, Libya. Geolines (Prague) 15:107–112

Németh K, Cronin SJ, Smith IEM, Flores JA (2012) Amplified hazard of small-volume monogenetic eruptions due to environmental controls, Orakei Basin, Auckland Volcanic field, New Zealand. Bull Volcanol 74(9):2121–2137

Novak O, Ritter JRR, Altherr R, Garasic V, Volker F, Kluge C, Kaspar T, Byrne GF, Sobolev SV, Fuchs K (1997) An integrated model for the deep structure of the Chyulu Hills volcanic field, Kenya. Tectonophysics 278(1–4):187–209

Pallister JS (1985) Reconnaissance geology of the Harrat Hutaymah quadrangle, sheet 26/42A, Kingdom of Saudi Arabia. Open-File Report—U.S. Geological Survey:82–82

Pallister JS, McCausland WA, Jonsson S, Lu Z, Zahran HM, El Hadidy S, Aburukbah A, Stewart ICF, Lundgren PR, White RA, Moufti MRH (2010) Broad accommodation of rift-related extension recorded by dyke intrusion in Saudi Arabia. Nat Geosci 3(10):705–712

Pankhurst RJ, Leat PT, Sruoga P, Rapela CW, Marquez M, Storey BC, Riley TR (1998) The Chon Aike province of Patagonia and related rocks in West Antarctica: A silicic large igneous province. J Volcanol Geoth Res 81(1–2):113–136

Petrone CM (2010) Relationship between monogenetic magmatism and stratovolcanoes in western Mexico: The role of low-pressure magmatic processes. Lithos 119(3–4):585–606

Prieto JLP (2013) Geositios, geomorfositios y geoparques: importancia, situación actual y perspectivas en México perspectives in Mexico. Inv Geogr 82:24–37

Rocha J, Brilha J, Henriques MH (2014) Assessment of the geological heritage of Cape Mondego Natural Monument (Central Portugal). Proc Geol Assoc 125(1):107–113

Runge MG, Bebbington MS, Cronin SJ, Lindsay JM, Kenedi CL, Moufti MRH (2014) Vents to events: determining an eruption event record from volcanic vent structures for the Harrat Rahat, Saudi Arabia. Bull Volc 76(3):1–16

Sakkas V, Meju MA, Khan MA, Haak V, Simpson F (2002) Magnetotelluric images of the crustal structure of Chyulu Hills volcanic field, Kenya. Tectonophysics 346(3–4):169–185

Schmincke H-U, Lorenz V, Seck HA (1983) The quaternary eifel volcanic fields. In: Fuchs K (ed) Plateau Uplift. Springer-Verlag, Berlin, Heidelberg, pp 139–151

Seghedi I, Helvaci C, Pecskay Z (2015) Composite volcanoes in the south-eastern part of Izmir-Balikesir Transfer Zone, Western Anatolia, Turkey. J Volcanol Geoth Res 291:72–85

Shaw JE, Baker JA, Kent AJR, Ibrahim KM, Menzies MA (2007) The geochemistry of the Arabian lithospheric mantle—A source for intraplate volcanism? J Petrol 48(8):1495–1512

Snyder GA, Taylor LA, Jerde EA, Sharkov Y, Laz'ko Y, Hanna S (1993) Petrogenesis of garnet pyroxenite and spinel peridotite xenoliths of the Tell-Danun alkali basalt volcano, Harrat as Shamah, Syria. Int Geol Rev 35(12):1104–1120

Spath A, Le Roex AP, Opiyo-Akech N (2001) Plume-lithosphere interaction and the origin of continental rift-related alkaline volcanism—The Chyulu Hills Volcanic Province, southern Kenya. J Petrol 42(4):765–787

Strba Lu, Rybar P, Balaz B, Molokac M, Hvizdak L, Krsak B, Lukac M, Muchova L, Tometzova D, Ferencikova J (2015) Geosite assessments: comparison of methods and results. Curr Issues Tour 18(5):496–510

Thornber CR (1990) Geologic map of Harrat Hutaymah, with petrologic classification and distribution of ultramafic inclusions, Saudi Arabia. In: U.S. Geological Survey, Reston, VA, Reston, VA, United States (USA), pp 0–1 sheet

Tomic N, Bozic S (2014) A modified geosite assessment model (M-GAM) and its application on the Lazar Canyon area (Serbia). Int J Env Res 8(4):1041–1052

Ukstins IA, Renne PR, Wolfenden E, Baker J, Ayalew D, Menzies M (2002) Matching conjugate volcanic rifted margins: Ar-40/Ar-39 chrono-stratigraphy of pre- and syn-rift bimodal flood volcanism in Ethiopia and Yemen. Earth Planet Sci Lett 198(3–4):289–306

Valentine GA, Cortes JA (2013) Time and space variations in magmatic and phreatomagmatic eruptive processes at Easy Chair (Lunar Crater Volcanic Field, Nevada, USA). Bull Volc 75(9):1–13

Vespermann D, Schmincke H-U (2000) Scoria cones and tuff rings. In: Sigurdsson H, Houghton BF, McNutt SR, Rymer H, Stix J (eds) Encyclopedia of volcanoes. Academic Press, San Diego, pp 683–694

Vujičić M, Vasiljević ĐA, Marković SB, Hose TA, Lukić T, Hadžić O, Janićević S (2011) Preliminary geosite assessment model (GAM) and its application on Fruška Gora Mountain, potential geotourism destination of Serbia. Acta Geogr Slov Geogr Zb 51(2):361–377

Wahab AA, Maaty MAA, Stuart FM, Awad H, Kafafy A (2014) The geology and geochronology of Al Wahbah maar crater, Harrat Kishb, Saudi Arabia. Quat Geochronol 21:70–76

Walker GPL (2000) Basaltic volcanoes and volcanic systems. In: Sigurdsson H, Houghton BF, McNutt SR, Rymer H, Stix J (eds) Encyclopedia of volcanoes. Academic Press, San Diego, pp 283–290

Wallace PJ, Carmichael ISE (1999) Quaternary volcanism near the Valley of Mexico: Implications for subduction zone magmatism and the effects of crustal thickness variations on primitive magma compositions. Contrib Miner Petrol 135(4):291–314

Wolff JA, Sumner JM (2000) Lava fountains and their products. In: Sigurdsson H, Houghton BF, McNutt SR, Rymer H, Stix J (eds) Encyclopedia of volcanoes. Academic Press, San Diego, pp 321–329

Woodland AB, Jugo PJ (2007) A complex magmatic system beneath the Deves volcanic field, Massif central, France: Evidence from clinopyroxene megacrysts. Contrib Miner Petrol 153(6):719–731

Yurur MT, Chorowicz J (1998) Recent volcanism, tectonics and plate kinematics near the junction of the African, Arabian and Anatolian plates in the eastern Mediterranean. J Volcanol Geoth Res 85(1–4):1–15

Zobin VM, Al-Amri AM, Fnais M (2013) Seismicity associated with active, new-born, and re-awakening basaltic volcanoes: Case review and the possible scenarios for the Harraat volcanic provinces, Saudi Arabia. Arabian J Geosci 6(2):529–541

Zunino M, Cavagna S, Clari P, Pavia G (2012) Two examples of the geoheritage potential of piedmont (NW Italy): Verrua Savoia for the present, Valle Ceppi for the future. Geoheritage 4(3):165–175

박준형, Cheong D (2012) A quantitative evaluation model for geoheritage. J Geol Soc Korea 48(2):163–178

MIX
Papier aus verantwortungsvollen Quellen
Paper from responsible sources
FSC® C105338

If you have any concerns about our products,
you can contact us on
ProductSafety@springernature.com

In case Publisher is established outside the EU,
the EU authorized representative is:
Springer Nature Customer Service Center GmbH
Europaplatz 3, 69115 Heidelberg, Germany

Printed by Libri Plureos GmbH
in Hamburg, Germany